Ku W9-ATT-463

Soil Bioventing
Principles and Practice

Soil Bioventing
Principles and Practice

Andrea Leeson
Robert E. Hinchee

With contributions by:
Bruce A. Alleman, Douglas C. Downey, Gregory
Headington, Jeffrey A. Kittel, Priti Kumar, Lt. Colonel
Ross N. Miller, Say Kee Ong, Gregory D. Sayles,
Lawrence Smith, Catherine M. Vogel

LEWIS PUBLISHERS

Boca Raton   New York   London   Tokyo

Acquiring Editor:	Neil Levine
Project Editor:	Suzanne Lassandro
Marketing Manager:	Greg Daurelle
Direct Marketing Manager:	Arline Massey
Cover design:	Dawn Boyd
PrePress:	Greg Cuciak
Manufacturing:	Sheri Schwartz

Library of Congress Cataloging-in-Publication Data

Leeson, Andrea, 1962–
    Soil bioventing : principles and practices / by Andrea Leeson
and Robert E. Hinchee : with contributions by Bruce A. Alleman ... [et at.]
      p. cm.
    Includes bibliographical references and index.
    ISBN 1-56670-126-0
    1. Soil bioventing. I. Hinchee, Robert E. II. Title.
TD878.6.L44 1996
628.5′5—dc20
                                       96-43809
                                        CIP

This book contains information obtained from authentic and highly regarded sources. Reprinted material is quoted with permission, and sources are indicated. A wide variety of references are listed. Reasonable efforts have been made to publish reliable data and information, but the author and the publisher cannot assume responsibility for the validity of all materials or for the consequences of their use.

Neither this book nor any part may be reproduced or transmitted in any form or by any means, electronic or mechanical, including photocopying, microfilming, and recording, or by any information storage or retrieval system, without prior permission in writing from the publisher.

The consent of CRC Press does not extend to copying for general distribution, for promotion, for creating new works, or for resale. Specific permission must be obtained in writing from CRC Press for such copying.

Direct all inquiries to CRC Press, Inc., 2000 Corporate Blvd., N.W., Boca Raton, Florida 33431.

© 1997 by CRC Press, Inc.
Lewis Publishers is an imprint of CRC Press

No claim to original U.S. Government works
International Standard Book Number 1-56670-126-0
Library of Congress Card Number   96-43809
Printed in the United States of America 1 2 3. 4 5 6 7 8 9 0
Printed on acid-free paper

To Alex

TABLE OF CONTENTS

i

LIST OF TABLES

LIST OF FIGURES

LIST OF EXAMPLES

ABBREVIATIONS AND ACRONYMS

AFB	Air Force Base
AFCEE	U.S. Air Force Center for Environmental Excellence
AFP	Air Force Plant
AFS	Air Force Station
AL/EQ	Armstrong Laboratory Environics Directorate
ANGB	Air National Guard Base
BTEX	benzene, toluene, ethylbenzene, and xylenes
cfm	cubic ft per minute
DNAPL	dense, nonaqueous-phase liquid
LNAPL	less dense, nonaqueous-phase liquid
NAS	Naval Air Station
NLIN	nonlinear regression procedure
PAHs	polycyclic aromatic hydrocarbons
PCBs	polychlorinated biphenyls
ppmv	parts per million, volume per volume
RD&A	Research, Development, and Acquisition
R_I	radius of influence
SAS	Statistical Analysis System
SVE	soil vacuum extraction
TCE	trichloroethylene
TKN	total Kjeldahl nitrogen
TPH	total petroleum hydrocarbon
TVH	total volatile hydrocarbon
UST	underground storage tank
U.S. EPA	U.S. Environmental Protection Agency
VOC	volatile organic carbon

SYMBOLS USED IN CALCULATIONS

C_S	quantity sorbed to the solid matrix
C_V	volumetric concentration in the vapor phase
C_{vsat}	saturated vapor concentration
C_W	volumetric concentration in the aqueous phase
E_a	activation energy
f_{oc}	organic carbon fraction
k	maximum rate of substrate utilization
k_d	endogenous respiration rate
k_o	baseline biodegradation rate

k_B	biodegradation rate
K_d	sorption coefficient
K_{ow}	octanol/water partition coefficient
K_S	Monod half-velocity constant
k_T	temperature-corrected biodegradation rate
MW	molecular weight
P_V	vapor pressure of pure contaminant at temperature T
Q	flowrate
R	gas constant
R_I	radius of influence
S	concentration of the primary substrate (contaminant)
s_x	solubility in water
t	time
T_{abs}	absolute temperature (°K)
χ	mole fraction
X	concentration of microorganisms
Y	cell yield

ACKNOWLEDGEMENTS

The authors would like to thank the following people for serving as peer reviewers for this document: R. Ryan Dupont, Ph.D., Utah State University; Jack van Eyk, Ph.D., Delft Geotechnic; Paul Johnson, Ph.D., Arizona State University, and Chi-Yuan Fan, U.S. EPA National Risk Management Laboratory.

The authors also would like to thank the following people for their help in finishing this document: Amanda Bush, Dr. Gordon Cobb, Rhonda Copley, Lynn Copley-Graves, Dean Foor, Jim Gibbs, Patrick Haas, Amy Householder, Wendy Huang, Gina Melaragno, Christine Peterson, Carol Young, and Dr. George Yu.

This document is a product of the bioventing research and development efforts sponsored by the U.S. Air Force Armstrong Laboratory, the Bioventing Initiative sponsored by the U.S. Air Force Center for Environmental Excellence (AFCEE) Technology Transfer Division, and the Bioremediation Field Initiative sponsored by the U.S. Environmental Protection Agency (U.S. EPA). The field research and data analysis was conducted by Battelle Memorial Institute and Parsons Engineering-Science.

CHAPTER 1

INTRODUCTION

Bioventing is the process of aerating soils to stimulate in situ biological activity and promote bioremediation. Bioventing typically is applied in situ to the vadose zone and is applicable to any chemical that can be aerobically biodegraded, but to date has been implemented primarily at petroleum-contaminated sites. Through the efforts of the U.S. Air Force Bioventing Initiative and the U.S. EPA Bioremediation Field Initiative, bioventing has been implemented at more than 150 sites and has emerged as one of the most cost-effective and efficient technologies currently available for vadose zone remediation of petroleum-contaminated sites. This document is a culmination of the experience gained from these sites and provides specific guidelines on the principles and practices of bioventing.

Much of the hydrocarbon residue at a fuel-contaminated site is found in the vadose zone soils, in the capillary fringe, and immediately below the water table (Figure 1-1). Seasonal water table fluctuations typically spread residues in the area immediately above and below the water table. Conventional physical treatment in the past involved pump-and-treat systems where groundwater was pumped out of the ground, treated, and either discharged or reinjected. Although useful for preventing continued migration of contaminants, these systems rarely achieved typical cleanup goals. Bioventing systems are designed to remove the contaminant source from the vadose zone, thereby preventing future and/or continued contamination of the groundwater.

A typical bioventing system is illustrated in Figure 1-2. Although bioventing is related to the process of soil vacuum extraction (SVE), the primary objectives of these two bioremediation technologies are different. Soil vacuum extraction is designed and operated to maximize the volatilization of low-molecular-weight compounds, with some biodegradation occurring. In contrast, bioventing is designed to maximize biodegradation of aerobically biodegradable compounds, regardless of their molecular weight, with some volatilization occurring. The major distinction between these technologies is that the objective of soil venting is to optimize removal by volatilization, while the objective of bioventing is to optimize biodegradation while minimizing volatilization and capital and utility costs. Although both technologies involve venting of air though the subsurface, the differences in objectives result in different design and operation of the remedial systems.

1

2

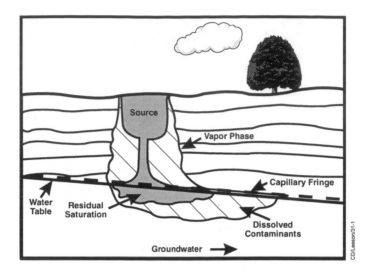

Figure 1-1. Hydrocarbon Distribution at a Typical Contaminated Site

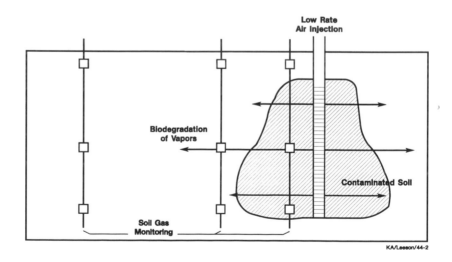

Figure 1-2. Schematic Diagram of a Typical Bioventing System

The results from bioventing research and development efforts and from the pilot-scale bioventing systems have been used to produce this two-volume manual. Although this design manual has been written based on extensive experience with petroleum hydrocarbons (and thus, many examples use this contaminant), the concepts here should be applicable to any aerobically biodegradable compound. The manual provides details on bioventing principles; site characterization; field treatability studies; system design, installation, and operation; process monitoring; site closure; and costs of implementation.

An overview of the development of bioventing, including development of the the U.S. Air Force Bioventing Initiative is provided as a basis for the data presented in this document. Data from U.S. Air Force Bioventing Initiative sites are used throughout this document to illustrate principles of bioventing as determined from field testing.

CHAPTER 2

DEVELOPMENT OF BIOVENTING

This chapter is intended to provide a framework for this document, describing the development and structure of the U.S. Air Force Bioventing Initiative and, ultimately, this document. This chapter provides an overview of bioventing, covering oxygen supply in situ — the dominant issue in the evolution of bioventing — early bioventing studies which led to development of the U.S. Air Force Bioventing Initiative, and the final structure of treatability studies and bioventing system design used for the U.S. Air Force Bioventing Initiative.

I. OXYGEN SUPPLY TO CONTAMINATED AREAS

One of the main driving forces behind the development of bioventing was the difficulty in delivering oxygen in situ. Many contaminants, especially the petroleum hydrocarbons found in fuels, are biodegradable if oxygen is available. Traditionally, enhanced bioreclamation processes used water to carry oxygen or an alternative electron acceptor to the contaminated zone. This was common whether the contamination was present in the groundwater or in the unsaturated zone. Media for adding oxygen to contaminated areas have included pure oxygen-sparged water, air-sparged water, hydrogen peroxide, and air.

In all cases where water is used, the solubility of oxygen is the limiting factor. At standard conditions, a maximum of 8 to 10 mg/L of oxygen can be obtained in water when aerated, while 40 to 50 mg/L can be obtained if sparged with pure oxygen, and up to 500 mg/L of oxygen theoretically can be supplied utilizing 1,000 mg/L of hydrogen peroxide. Using the stoichiometric equation shown as Equation 2-1[1], the quantity of water which must be delivered to provide sufficient oxygen for biodegradation can be calculated.

$$C_6H_{14} + 9.5O_2 \rightarrow 6CO_2 + 7H_2O \qquad (2\text{-}1)$$

---

[1] Refer to Chapter 3 for development of this equation.

An example of calculating the mass of water that must be delivered for hydrocarbon degradation is shown in Example 2-1. Table 2-1 summarizes oxygen requirements based on the supplied form of oxygen.

Example 2-1. Calculation of Air-Saturated Water Mass That Must Be Delivered to Degrade Hydrocarbons: Based on Equation 2-1, the stoichiometric molar ratio of hydrocarbon to oxygen is 1:9.5. Or, to degrade one mole of hydrocarbons, 9.5 moles of oxygen must be consumed. On a mass basis:

$$\frac{1 \text{ mole } C_6H_{14}}{9.5 \text{ moles } O_2} \times \frac{1 \text{ mole } O_2}{32 \text{ g } O_2} \times \frac{86 \text{ g } C_6H_{14}}{1 \text{ mole } C_6H_{14}} = \frac{86 \text{ g } C_6H_{14}}{304 \text{ g } O_2} = \frac{1 \text{ g } C_6H_{14}}{3.5 \text{ g } O_2}$$

Given an average concentration of 9 mg/L oxygen dissolved in water, the amount of air-saturated water that must be delivered to degrade 1 g hydrocarbon is calculated as follows:

$$\frac{3.5 \text{ g } O_2 \text{ required}}{\frac{9 \text{ mg } O_2}{1 \text{ L } H_2O} \times \frac{1 \text{ g}}{1,000 \text{ mg}}} = \frac{390 \text{ L } H_2O}{1 \text{ g } C_6H_{14}}$$

or, to degrade 1 lb:

$$\frac{390 \text{ L } H_2O}{1 \text{ g } C_6H_{14}} \times \frac{1 \text{ gallon}}{3.8 \text{ L}} \times \frac{1,000 \text{ g}}{2.2 \text{ lb}} = \frac{47,000 \text{ gallons } H_2O}{1 \text{ lb } C_6H_{14}}$$

Due to the low aqueous solubility of oxygen, hydrogen peroxide has been tested as an oxygen source in laboratory studies and at several field sites (Hinchee et al., 1991a; Aggarwal et al., 1991; Morgan and Watkinson, 1992). As shown in Table 2-1, if 500 mg/L of dissolved oxygen can be supplied via hydrogen peroxide, the mass of water that must be delivered is reduced by more than an order of magnitude. Initially, these calculations made the use of hydrogen peroxide appear to be an attractive alternative to injecting air-saturated water.

Hydrogen peroxide is miscible in water and decomposes to release water and oxygen as shown in Equation 2-2:

$$H_2O_2 \rightarrow H_2O + \frac{1}{2}O_2 \qquad (2-2)$$

Many substances commonly present in groundwater and soils act as catalysts for the decomposition of peroxide. Important among these are aqueous species of iron and copper, and the enzyme catalase (Schumb et al., 1955), which has significant activity in situ (Spain et al., 1989). If the rate of oxygen formation from hydrogen peroxide decomposition exceeds the rate of microbial oxygen utilization, gaseous oxygen may form due to its limited aqueous solubility. Gaseous oxygen may form bubbles that may not be transported efficiently in groundwater, resulting in ineffective oxygen delivery.

Table 2-1
Oxygen Requirements Based on Supplied Form of Oxygen

Oxygen Form	Oxygen Concentration in H_2O	Volume to Degrade 1 lb Hydrocarbon
Air-saturated H_2O	8 to 10 mg/L	47,000 gallons (180,000 L)
Oxygen-saturated H_2O	40 to 50 mg/L	11,000 gallons (42,000 L)
Hydrogen peroxide	Up to 500 mg/L	1,600 gallons (6,100 L)
Air	NA (21% vol/vol in air)	170 ft^3 (4,800 L)

Phosphate is commonly used in nutrient formulations in an effort to decrease the rate of peroxide decomposition in groundwater applications (Britton, 1985). However, the effectiveness of phosphate addition in stabilizing peroxide injected into an aquifer has not been well established and conflicting results have been reported by different researchers (American Petroleum Institute, 1987; Brown et al., 1984; Downey et al., 1988; Huling et al., 1990; Morgan and Watkinson, 1992).

A field experiment was conducted by Hinchee et al. (1991a) to examine the effectiveness of hydrogen peroxide as an oxygen source for in situ biodegradation. The study was performed at a JP-4 jet fuel-contaminated site at Eglin AFB, Florida. Site soils consisted of fine- to coarse-grained quartz sand with groundwater at a depth of 2 to 6 ft (0.61 to 1.8 m). Previous studies by Downey ct al. (1988) and Hinchee et al. (1989b) at the same site had shown that rapid decomposition of hydrogen peroxide occurred, even with the addition of phosphate as a peroxide stabilizer. In subsequent studies, hydrogen peroxide was injected at a concentration of 300 mg/L both with and without the addition of a phosphate-containing nutrient solution. As in previous studies, hydrogen peroxide decomposition was rapid, resulting in poor distribution of oxygen in groundwater. Addition of the phosphate-containing nutrient solution did not appear to improve hydrogen peroxide stability.

Other attempts have been made using hydrogen peroxide as an oxygen source. Although results indicate better hydrogen peroxide stability than achieved by Hinchee et al. (1989a), it was concluded that most of the hydrogen peroxide decomposed rapidly (Huling et al., 1990). Some degradation of aromatic hydrocarbons appears to have occurred; however, no change in total hydrocarbon contamination levels was detected in the soils (Ward, 1988).

In contrast to hydrogen peroxide use, when air is used as an oxygen source in unsaturated soil, 170 ft^3 (4,800 L) of air must be delivered to provide the minimum oxygen required to degrade 1 lb (0.45 kg) of hydrocarbon (Table 2-1). Since costs associated with water-based delivery of oxygen can be relatively high, the use of gas-phase delivery results in a significant reduction in the cost associated with supplying oxygen[1].

An additional advantage of using a gas-phase process is that gases have greater diffusivity than liquids. At many sites, geologic heterogeneities cause fluid that is pumped through the formation to be channelled into the more-permeable pathways (e.g., in an alluvial soil with interbedded sand and clay, all of the fluid flow initially takes place in the sand). As a result, oxygen must be delivered to the less-permeable clay lenses through diffusion. In a gaseous system (as found in unsaturated soils), this diffusion can be expected to take place at rates at least three orders of magnitude greater than rates in a liquid system (as is found in saturated soils). Although it is not realistic to expect diffusion to aid significantly in water-based bioreclamation, diffusion of oxygen in a gas-phase system is a significant mechanism for oxygen delivery to less-permeable zones.

Given the advantages of using air rather than water as the oxygen source, several investigators began exploring the feasibility of an air-based oxygen supply system as a remedial option. A summary of the results of these investigations is presented in the following section.

II. BIOVENTING RESEARCH AND DEVELOPMENT

Figure 2-1 provides a historical perspective of bioventing research and development. To the authors' knowledge, the first documented evidence of unsaturated zone biodegradation resulting from forced aeration was reported by the Texas Research Institute, Inc., in a 1980 study for the American Petroleum Institute. A large-scale model experiment was conducted to test the effectiveness of a surfactant treatment to enhance the recovery of spilled gasoline. The experiment accounted for only 8 gallons (30 L) of the 65 gallons (250 L) originally spilled and raised questions about the fate of the gasoline. Subsequently, a column study was conducted to determine a diffusion coefficient for soil venting. This column study evolved into a biodegradation study in which it was concluded that as much as 38% of the fuel hydrocarbons were biologically mineralized. Researchers concluded that venting not only would remove gasoline by physical means, but also would enhance microbial activity and promote biodegradation of the gasoline (Texas Research Institute, 1980; 1984).

---

[1] Refer to Chapter 9 for a comparison of costs associated with hydrogen peroxide use versus air (bioventing).

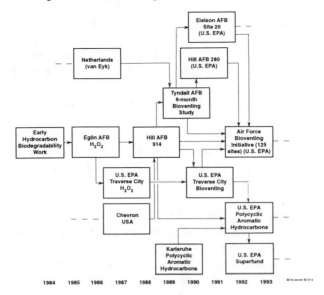

Figure 2-1. Historical Perspective of the Development of Bioventing

To the authors' knowledge, the first actual field-scale bioventing experiments were conducted by Jack van Eyk for Shell Research. In 1982, at van Eyk's direction, the Shell Laboratory in Amsterdam, The Netherlands initiated a series of experiments to investigate the effectiveness of bioventing for treating hydrocarbon-contaminated soils. These studies were reported in a series of papers (Anonymous, 1986; Staatsuitgeverij, 1986; van Eyk and Vreeken, 1988; 1989a; and 1989b).

Wilson and Ward (1986) suggested that using air as a carrier for oxygen could be 1,000 times more efficient than using water, especially in deep, hard-to-flood unsaturated zones. They made the connection between oxygen supply via soil venting and biodegradation by observing that "soil venting uses the same principle to remove volatile components of the hydrocarbon." In a general overview of the soil venting process, Bennedsen et al. (1987) concluded that soil venting provides large quantities of oxygen to the unsaturated zone, possibly stimulating aerobic degradation. They suggested that water and nutrients also would be required for significant degradation and encouraged additional investigation into this area.

Biodegradation enhanced by soil venting has been observed at several field sites. Investigators claim that at a soil venting site for remediation of gasoline-contaminated soil, significant biodegradation occurred (measured by a temperature rise) when air was supplied. Investigators pumped pulses of air through a pile of excavated soil and observed a consistent rise in temperature, which they attributed to biodegradation. They claimed that the pile was cleaned up during the summer primarily by biodegradation (Conner, 1989). However, they did not control for natural volatilization from the aboveground pile, and not enough data were published to critically review their biodegradation claim.

Researchers at Traverse City, Michigan, observed a decrease in the toluene concentration in unsaturated zone soil gas, which they measured as an indicator of fuel contamination in the unsaturated zone. They assumed that advection had not occurred and attributed the toluene loss to biodegradation. The investigators concluded that because toluene concentrations decayed near the oxygenated ground surface, soil venting is an attractive remediation alternative for biodegrading light volatile hydrocarbon spills (Ostendorf and Kampbell, 1989).

The U.S. Air Force initiated its research and development program in bioventing in 1988 with a study at Site 914, Hill AFB, Utah, discussed in Chapter 4. This site initially was operated as a soil vapor extraction unit, but was modified to a bioventing system after 9 months of operation because there was evidence of biodegradation and in an effort to reduce costs by reducing off-gas. Moisture and nutrient addition were studied at this site; however, while moisture addition appeared to improve biodegradation, nutrient addition did not. Final soil sampling demonstrated that benzene, toluene, ethylbenzene, and xylenes (BTEX) and total petroleum hydrocarbon (TPH) levels were reduced to below regulatory levels, and this site became the first Air Force site that was closed through in situ bioremediation. During this study, it became apparent that bioventing had great potential for remediating JP-4 jet fuel-contaminated soils. It also was apparent that additional research would be needed before the technology could be applied routinely in the field.

Following the Site 914, Hill AFB study, a more controlled bioventing study was completed at Tyndall AFB, Florida, discussed in Chapter 4. This study was designed to monitor specific process variables and the subsequent effect on biodegradation of hydrocarbons. Several important findings resulted from this work, including the effect of air flowrates on removal by biodegradation and volatilization, the effect of temperature on biodegradation rates, the lack of microbial stimulation from the addition of moisture and nutrients, and the importance of natural nitrogen supply through nitrogen fixation. In addition, initial and final contaminant measurements showed over 90% removal of BTEX. Although this study was short-term, it illustrated the effectiveness of bioventing.

The studies conducted at Hill and Tyndall AFBs provided valuable information on bioventing. However, it was apparent that long-term, controlled bioventing studies were necessary to fully evaluate and optimize the technology. In 1991, long-term bioventing studies were initiated at Site 280, Hill AFB, Utah and at Site 20, Eielson AFB, Alaska, discussed in

Chapter 4. These studies were joint efforts between the U.S. EPA and the U.S. Air Force Environics Directorate of the Armstrong Laboratory. These studies have involved intensive monitoring of several process variables, including the effect of soil temperature on biodegradation rates, surface emission analyses, and optimization of flowrate.

Based on the success of these previous studies, in 1992, AFCEE initiated the U.S. Air Force Bioventing Initiative where pilot-scale bioventing systems were installed at 125 contaminated sites located throughout the continental United States and in Hawaii, Alaska, and Johnston Atoll (Figure 2-2). The sites varied dramatically in climatic and geologic conditions. Contaminants typically were petroleum hydrocarbons from JP-4 jet fuel, heating oils, waste oils, gasoline, and/or diesel; however, some fire training areas also were studied where significant concentrations of solvents were present. This manual is a product of this study and represents the culmination of data collected from these sites and other projects.

In addition to these studies, other bioventing studies have been conducted by several researchers. A summary of select sites where bioventing has been applied is shown in Table 2-2. The scale of application and contaminant type is given, as well as the biodegradation rate, if known. The studies listed in Table 2-2 are limited to those where the study was conducted in situ, where no inoculum was added to site soils, and flowrates were optimized for biodegradation, not volatilization. It is important to distinguish between bioventing and SVE systems. Bioventing systems operate at flowrates optimized for biodegradation not volatilization, although some volatilization may occur. SVE systems operate at flowrates optimized for volatilization, although some biodegradation may occur. Therefore, flowrates and configuration of the two systems are significantly different.

The following section describes the basic structure for field studies conducted as part of the U.S. Air Force Bioventing Initiative. Data from these studies were used to generate this two-volume document.

III. STRUCTURE OF U.S. AIR FORCE BIOVENTING INITIATIVE FIELD TREATABILITY STUDIES AND BIOVENTING SYSTEM DESIGN

The design of the field treatability studies and final bioventing system was developed based on experience at previous studies at Hill, Tyndall, and Eielson AFBs. The *Test Plan and Technical Protocol for a Field Treatability Test for Bioventing* (Hinchee et al., 1992) was written to standardize all field methods from treatability tests to well installations. The document allowed for collection of consistent data from 125 sites, which provided a strong database for evaluating bioventing potential. At

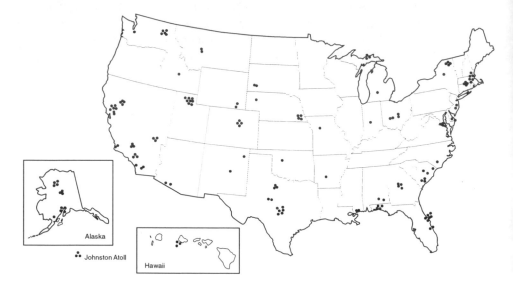

Figure 2-2. Locations of Bioventing Initiative Sites

all sites, the following activities were conducted:

- site characterization, including a small-scale soil gas survey and collection of initial soil and soil gas samples for analysis of BTEX, TPH, and soil physicochemical characteristics;
- field treatability studies, including an in situ respiration test and a soil gas permeability test;
- identification of a background, uncontaminated area for comparison with the contaminated area of background respiration rates and nutrient levels;
- installation of a blower for 1-year of operation (typically configured for air injection), if results of field treatability studies were positive;
- 6-month and 1-year in situ respiration tests at sites where a blower had been installed; and
- collection of final soil and soil gas samples for analyses of BTEX and TPH.

Of particular significance were the use of the in situ respiration test to measure microbial activity and the use of air injection instead of extraction for air delivery.

The in situ respiration test was developed to rapidly measure aerobic biodegradation rates in situ at discrete locations. Biodegradation rates calculated from the in situ respiration test are useful (1) for assessing the

Table 2-2
Summary of Reported In Situ Respiration Rates and Bioventing Data

Site	Scale of Application	Contaminant	In Situ Respiration Rates (mg/kg-day unless marked)	Reference
Albemarle County, VA	Pilot scale	Acetone, toluene, benzene, naphthalene	1.5 – 26	Leeson et al., 1994
Eielson AFB, AK	Pilot scale	JP-4 jet fuel	0.82 – 8.2	Hinchee & Ong, 1992 Leeson et al., 1995 Sayles et al., 1994a
Fallon NAS, NE	In situ respiration test	JP-5 jet fuel	4.2	Hinchee et al., 1991b
Galena AFS, AK, Saddle Tank Farm	Pilot scale	Diesel	11 – 30	Ong et al., 1994
Hill AFB, Utah, Site 914	Full scale, 2 years	JP-4 jet fuel	up to 8.5	Hinchee et al., 1991a
Hill AFB, Utah, Site 280	Full scale	JP-4 jet fuel	0.27 (site average)	Battelle, 1994 Sayles et al., 1994b
Eglin AFB, FL	Full Scale	Gasoline	4.0	Downey et al., 1994
Kenai, Alaska, Site 1-33	Pilot scale	Crude oil, petroleum	2.7 – 25	Hinchee, unpublished data
Kenai, Alaska, Site 3-9	Pilot scale	Crude oil	0.64 – 12	Hinchee, unpublished data
Massachusetts	Full scale	Gasoline	Not measured	Brown & Crosbie, 1994
Minnesota	Full scale	Gasoline	15, 4.9, 3.1, 0.20	Newman et al., 1993
The Netherlands	Full scale	Gasoline	570 kg hydrocarbon removed during 2-yr study	van Eyk, 1994
The Netherlands	Undefined	Undefined	1.6 – 4.2	Urlings et al., 1990

Location	Scale/Test	Fuel	Rate	Reference
The Netherlands	Field pilot, 1 year	Diesel	6.9	van Eyk & Vreeken, 1989b
Patuxent River NAS, MD	In situ respiration test	JP-5 jet fuel	2.6	Hinchee et al., 1991b
Prudhoe Bay	Pilot scale	Diesel	8.6 — 11	Ong et al., 1994
St. Louis Park, MN, Reilly Tar Site	Pilot scale	PAH	0.55—2.2 mg PAH/kg-day	Alleman et al., 1995
Seattle, WA	Full Scale	Diesel	6.0	Baker et al., 1993
Southern CA	Full scale	Gasoline, hydraulic oil	0.14	Zachary & Everett, 1993
Tinker AFB, OK	In situ respiration test	JP-4 and mixed fuels	2.3 — 15	Hinchee & Smith, 1991
Tyndall AFB, FL	Field pilot, 1 year and in situ respiration test	JP-4 jet fuel	1.6 — 16	Miller, 1990; Hinchee et al., 1991b
Undefined	Full scale	Gasoline and diesel	50 kg/(well day)	Ely & Heffner, 1988
Undefined	Full scale	Diesel	100 kg/(well day)	Ely & Heffner, 1988
Undefined	Full scale	Fuel oil	60 kg/(well day)	Ely & Heffner, 1988
Valdez, Site A	Pilot scale	Crude oil	8.7 — 16.0	Foor and Hinchee, 1993

potential application of bioremediation at a given site, (2) for estimating the time required for remediation at a given site, and (3) for providing a measurement tool for evaluating the effects of various environmental parameters on microbial activity and ultimately on bioventing performance. The actual effect of individual parameters on microbial activity is difficult to assess in the field due to interference and interactions among these parameters.   The in situ respiration test integrates all factors to simply assess whether the microorganisms are metabolizing the fuel.   Data from the in situ respiration test and site measurements were used to conduct a statistical analysis of the observed effects of the site measurements on microbial activity in the field.   The statistical analysis was constructed to account for parameter interactions.   These results are discussed in detail in Appendix A.

Also of note is that 120 of the 125 bioventing systems installed were configured for air injection.   Prior to the bioventing studies conducted at Hill (Site 280) and Eielson AFBs, bioventing systems typically were operated in the extraction configuration, similar to SVE systems. However, research at Hill and Eielson AFBs demonstrated that air injection is a feasible and more efficient alternative to air extraction, resulting in a greater proportion of hydrocarbon biodegradation rather than volatilization and reduced air emissions[1].   Therefore, the air injection configuration was selected for the basic bioventing system at U.S. Air Force Bioventing Initiative sites.

The results generated from the U.S. Air Force Bioventing Initiative are summarized in detail in Chapter 5 and are used to illustrate the basic principles of bioventing and microbial processes discussed in Chapter 3. The design guidelines presented in this manual have culminated primarily from the experience of installing and operating the 125 U.S. Air Force Bioventing Initiative sites.   These design guidelines represent the basic bioventing system, which is applicable to the majority of sites suitable for bioventing.   The following chapter addresses emerging techniques for modifications to the basic bioventing system described in this document for sites that are not amenable to standard bioventing methods.

---

[1]   Refer to Chapter 6 for a discussion of air injection versus extraction.

CHAPTER 3

PRINCIPLES OF BIOVENTING

In this chapter, basic principles fundamental to the bioventing process are discussed to provide a clear understanding of the many physical, chemical, and biological processes that impact the ultimate feasibility of bioventing. Recognizing the significance of these different processes will lead to more efficient bioventing design and operation. Specific topics to be considered in this chapter include:

- soil gas permeability, contaminant diffusion and distribution, and zone of oxygen influence;
- subsurface distribution of an immiscible liquid;
- environmental factors which affect microbial processes, such as electron acceptor conditions, moisture content, pH, temperature, nutrient supply, contaminant concentration, and bioavailability;
- compounds targeted for removal through bioventing; and
- BTEX versus TPH removal during petroleum bioventing.

I. PHYSICAL PROCESSES AFFECTING BIOVENTING

Primarily four physical parameters are significant to the bioventing process. These include soil gas permeability, contaminant diffusion in soil, contaminant distribution, and zone of oxygen influence. Each of these parameters is discussed in the following sections.

A. SOIL GAS PERMEABILITY
Assuming contaminants are present that are amenable to bioventing, geology probably is the most important site characteristic for a successful bioventing application. Soils must be sufficiently permeable to allow movement of enough soil gas to provide adequate oxygen for biodegradation, on the order of 0.25 to 0.5 pore volumes per day.

Soil gas permeability is a function of both soil structure and particle size, as well as of soil moisture content. Typically, permeability in excess of 0.1 darcy is adequate for sufficient air exchange. Below this level, bioventing certainly is possible, but field testing may be required to establish feasibility.

When the soil gas permeability falls below approximately 0.01 darcy, soil gas flow is primarily through either secondary porosity (such as

17

fractures) or through any more permeable strata that may be present (such as thin sand lenses). Therefore, the feasibility of bioventing in low-permeability soils is a function of the distribution of flow paths and diffusion of air to and from the flow paths within the contaminated area. Bioventing has been successful in some low-permeability soils, such as a silty clay site at Fallon Naval Air Station (NAS), Nevada (Kittel et al., 1995), a clayey site at Beale AFB, California (Phelps et al., 1995), a silty site at Eielson AFB, Alaska (Leeson et al., 1995), and a silty clay site in Albemarle County, Virginia (Leeson et al., 1994), and at many U.S. Air Force Bioventing Initiative sites.

In a soil that is of a reasonable permeability, a minimum separation of 2 to 4 ft (0.61 to 1.2 m) between vertical and horizontal flow paths and contaminant may still result in successful treatment due to oxygen diffusion. However, the degree of treatment will be very site-specific.

Grain size analysis was conducted on several samples from each site in the U.S. Air Force Bioventing Initiative. The relative distribution of fine-grained soils is illustrated in Figure 3-1. Approximately 50% of the sites tested contained greater than 50% clay and silt fractions. Sufficient soil gas permeability was demonstrated at many sites with silt and clay contents exceeding 80% by weight.

Oxygen distribution generally was adequate in soils where permeability values exceeded 0.1 darcy, with oxygen detected at ambient levels in all installed monitoring points. Few sites had a soil gas permeability less than 0.1 darcy; therefore, data for analysis are limited. The greatest limitation to bioventing at U.S. Air Force Bioventing Initiative sites was excessive soil moisture. A combination of high soil moisture content and fine-grained soils made bioventing impractical at only three of the 125 test sites.

In general, our calculated soil gas permeability values have exceeded suggested literature values reported in Johnson et al. (1990) for silt and clay soils. This is likely due to the heterogeneous nature of most soils, which contain lenses of more permeable material or fractures which aid in air distribution.

B. CONTAMINANT DISTRIBUTION

Because bioventing is in essence an air delivery system designed to efficiently provide sufficient oxygen to contaminated soils, it is important to have a clear understanding of subsurface contaminant distribution. Many of the sites at which bioventing can be applied are contaminated with immiscible liquids, such as petroleum hydrocarbons. When a fuel release occurs, the contaminants may be present in any or all of four phases in the geologic media:

- sorbed to the soils in the vadose zone;
- in the vapor phase in the vadose zone;

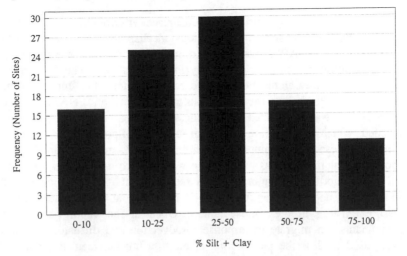

Figure 3-1. **Distribution of Silt- and Clay-Sized Particles at Bioventing Initiative Sites**

- in free-phase form floating on the water table or as residual saturation in the vadose zone; and/or
- in the aqueous phase dissolved in pore water in the vadose zone or dissolved in the groundwater.

Of the four phases, dissolved petroleum contaminants in the groundwater frequently are considered to be of greatest concern due to the risk of humans being exposed to contaminants through drinking water. However, the free-phase and sorbed-phase hydrocarbons act as feedstocks for groundwater contamination, so any remedial technology aimed at reducing groundwater contamination must address these sources of contamination. Also, hydrocarbons in the vadose zone can produce a volatile organic carbon (VOC) threat in subsurface buildings or structures.

Immiscible liquids are classified as less dense, nonaqueous-phase liquids (LNAPLs) if their density is less than water or dense, nonaqueous-phase liquids (DNAPLs) if their density is greater than water. In general, most petroleum hydrocarbons, such as gasoline, are LNAPLs, whereas most chlorinated solvents, such as trichloroethylene (TCE), are DNAPLs. Due to these differences in densities, subsurface spills of LNAPLs and DNAPLs will behave differently at a given site, with LNAPLs distributed primarily in the vadose zone and DNAPLs distributed in both the unsaturated and saturated zone. Given that bioventing is primarily a vadose zone treatment process, this discussion will focus on the behavior of LNAPLs.

When a large enough fuel spill occurs, the fuel is retained within approximately 10 to 20% of the pore volume of the soil and eventually may come to rest on the water table. Contaminants then partition among the various phases existing within the subsurface environment. Fluids can move through the subsurface via various mechanisms, such as advection and diffusion. LNAPLs are likely to migrate through the vadose zone relatively uniformly until the capillary fringe is reached. The LNAPL will then spread laterally along the saturated zone. Water table fluctuation may result in LNAPL below the water table; however, an LNAPL will not permeate the water-saturated zone unless a critical capillary pressure is exceeded, which is a function of the porous-medium pore sizes.

In the vadose zone, components of the LNAPL may partition into the vapor phase or the aqueous phase (pore water), sorb onto solids, or remain in the free product. Contaminants in free product may partition into the vapor phase, depending on their vapor pressures at the temperature and pressure existing in the vadose zone. Once in the vapor phase, these contaminants can migrate in response to advection and diffusion. Raoult's law is used to describe partitioning at equilibrium between an immiscible phase and a vapor phase:

$$C_V = \chi \; C_{vsat} \qquad (3\text{-}1)$$

where: C_V = volumetric concentration of the contaminant (x) in the vapor phase (g_x/L_{vapor});

χ = mole fraction of the contaminant (dimensionless); and

C_{vsat} = saturated vapor concentration of the contaminant (g_x/L_{vapor}).

C_{vsat} is further defined as:

$$C_{vsat} = \frac{(MW_x) \; P_V}{R T_{abs}} \qquad (3\text{-}2)$$

where: MW_x = molecular weight of the contaminant ($g_x/mole_x$);

P_V = vapor pressure of pure contaminant at temperature T (atm);

R = gas constant (L-atm/mole-°K); and

T_{abs} = absolute temperature (°K).

Free product in contact with groundwater may leach contaminants into the groundwater or contaminants may dissolve into pore water in the vadose zone, depending on the solubility of specific components. Once in the groundwater, contaminants can migrate through the subsurface in response to a gradient in the aqueous-phase total potential (i.e., advection) or by a difference in the aqueous-phase chemical concentrations. The equilibrium relationship between the aqueous and the immiscible phases is described as:

$$C_W = \chi \, s_x \qquad (3\text{-}3)$$

where: C_W = volumetric concentration of contaminant x in the aqueous phase ($g_x/L_{aqueous}$);

s_x = solubility of pure contaminant x in water (g_x/L_{water}).

Sorption of contaminants is a complex process involving several different phenomena including coulomb forces, London-van der Waals forces, hydrogen bonding, ligand exchange, dipole-dipole forces, dipole-induced dipole forces, and hydrophobic forces (Wiedemeier et al., 1995). In the case of hydrocarbons, due to their nonpolar nature, sorption most often occurs through hydrophobic bonding to organic matter. Hydrophobic bonding often is a dominant factor influencing the fate of organic chemicals in the subsurface (DeVinny et al., 1990). The degree of sorption generally is empirically related by the organic content of the soil and by the octanol-water partition coefficient of a particular compound.

Sorption isotherms generally follow one of three shapes: Langmuir, Freundlich, or linear (Figure 3-2). The Langmuir isotherm describes the sorbed contaminant concentration as increasing linearly with concentration then leveling off as the number of sites available for sorption are filled. This isotherm accurately describes the situation at or near the contaminant source where concentrations are high. The Freundlich isotherm assumes an infinite number of sorption sites, which would accurately describe an area some distance from the contaminant source where concentrations are dilute. The mathematical expression contains a chemical-specific coefficient that may alter the linearity of the isotherm. The linear isotherm is relatively simple and is valid for dissolved compounds at less than one-half of their solubility (Lyman et al., 1992). This isotherm is typically valid to describe hydrocarbon sorption.

The linear isotherm is expressed mathematically as:

$$C_s = K_d \, C_W \qquad (3\text{-}4)$$

where: C_s = quantity of contaminant x sorbed to the solid matrix (g_x/g_{soil});

K_d = sorption coefficient ($L_{aqueous}/g_{soil}$).

The sorption coefficient may be determined experimentally, estimated based on values published in the literature, or estimated using the octanol/water partition coefficient (K_{ow}) and the organic carbon fraction (f_{oc}) of the soil. The sorption coefficient can be estimated using the following mathematical expression:

$$K_d = K_{ow} \, f_{oc} \qquad (3\text{-}5)$$

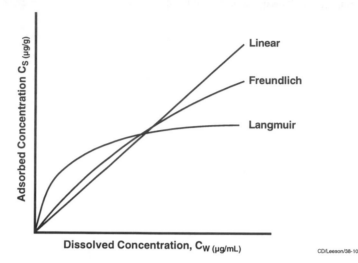

Figure 3-2. Sorption Isotherms

Some values for K_{ow} are provided in Table 3-1.

In practice, at equilibrium, the concentration of most petroleum hydrocarbon compounds of interest in the aqueous or vapor phases is driven by the immiscible phase, if present, and the sorbed phase, if the immiscible phase is not present. If no immiscible phase is present, and all sorption sites on the solid soil matrix are not occupied[1], the vapor or aqueous phase concentration is a function of the sorbed concentration. This relationship is illustrated in Figure 3-3.

This relationship typically follows a Langmuir type curve. If the concentration in the soil is in excess of the sorption capacity of the soil[2], the aqueous-phase and vapor-phase concentrations are Raoult's law-driven and are independent of the hydrocarbon concentration in the soil. This is an important concept in attempting to interpret soil gas or groundwater data. For example, in a sandy site at which free product has been detected, the highest soil hydrocarbon concentrations may exceed 25,000 mg/kg. Yet 99% remediation to 250 mg/kg may not affect the equilibrium soil gas or groundwater hydrocarbon concentrations.

In terms of contaminant distribution, difficulties in applying bioventing arise when significant quantities of the contaminant are in the capillary fringe or below the water table due to groundwater fluctuations. Treatment of the capillary fringe is possible and screening of venting wells below the

---

[1] In most soils, this is probably at a concentration of less than 100 to 1,000 mg/kg.

[2] In most soils, this is probably at a concentration greater than 100 to 1,000 mg/kg.

Table 3-1
Values for Key Properties of Select Petroleum Hydrocarbons

Compound	K_{ow}	Solubility (mg/L)	Vapor Pressure (mm Hg)[1]
Benzene	131.82	1,750[2]	75
Ethylbenzene	1,349	152[2]	10[79°F]
Heptane		50	40
Hexane		20	150
Toluene	489.9	537[3]	20[65°F]
o-xylene	891	152[2]	7
m-xylene	1,585	158[2]	9
p-xylene	1,513.6	198[2]	9

[1] Vapor pressure at 68°F unless noted.   [2] Calculated at 20°C.   [3] Calculated at 20°C.

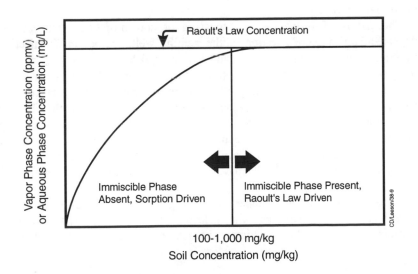

Figure 3-3. Relationship Between Sorbed Contaminant Concentration and Vapor- or Aqueous-Phase Concentrations

water table is recommended to ensure treatment of this area[1]. However, the ability of bioventing to aerate the capillary fringe and underlying water table has not been evaluated. Limited oxygenation is anticipated due to water-filled pore space. If significant contamination exists below the water table, dewatering should be considered as a means of exposing any contaminated soil to injected air. Alternatively, a combination of air sparging and bioventing may provide more efficient air delivery to the capillary fringe; however, air sparging has not been well-documented and many parameters are still unknown concerning its applicability and effectiveness.

C. OXYGEN RADIUS OF INFLUENCE

An estimate of the oxygen radius of influence (R_I) of venting wells is an important element of a full-scale bioventing design. This measurement is used to design full-scale systems, specifically to space venting wells, to size blower equipment, and to ensure that the entire site receives a supply of oxygen-rich air to sustain in situ biodegradation.

The radius of oxygen influence is defined as the radius to which oxygen has to be supplied to sustain maximal biodegradation. This definition of radius of influence is different than is typically used for SVE, where radius of influence is defined as the maximum distance from the air extraction or injection well where vacuum or pressure (soil gas movement) occurs. The oxygen radius of influence is a function of both air flowrates and oxygen utilization rates, and therefore depends on site geology, well design, and microbial activity.

The radius of influence is a function of soil properties, but also is dependent on the configuration of the venting well, extraction or injection flowrates, and microbial activity, and it is altered by soil stratification. In soils with less-permeable lenses adjacent to more-permeable soils, injection into the permeable layer will produce a greater radius of influence than could be achieved in homogeneous soils. On sites with shallow contamination, the radius of influence also may be increased by impermeable surface barriers such as asphalt or concrete. Frequently, however, paved surfaces do not act as vapor barriers. Without a tight seal to the native soil surface[2], the pavement will not significantly impact soil gas flow.

Microbial activity will impact the oxygen radius of influence. As microbial activity increases, the effective treated area will decrease. Therefore, it is desirable to estimate the oxygen radius of influence at times

---

[1] Refer to Chapter 6 for a discussion of vent well construction.

[2] It is the authors' experience that at most sites, this seal does not occur.

of peak microbial activity and to design the bioventing system based on these measurements.

II. MICROBIAL PROCESSES AFFECTING BIOVENTING

Biological treatment approaches rely on organisms to destroy or reduce the toxicity of contaminants. The advantages of chemical and physical treatment approaches generally are outweighed by the ability of microorganisms to mineralize contaminants, thereby eliminating the process of transferring contaminants from one medium (i.e., soil and soil vapor) into another (i.e., activated carbon) that will still require treatment. In addition, through microbial processes, it is possible to treat large areas relatively inexpensively and with relatively noninvasive techniques. This section discusses kinetics of microbial metabolism and environmental parameters which affect the microbial processes bioventing is dependent upon, thereby potentially affecting the efficacy of bioventing.

A. MICROBIAL KINETICS

In biological processes, microorganisms degrade organic compounds either directly to obtain carbon and/or energy, or fortuitously in a cometabolic process with no significant benefit to the microorganism. In the case of bioventing of petroleum hydrocarbons, microorganisms are capable of metabolizing many of the fuel components as a carbon and energy source. Because the microorganisms are stimulated in situ, the microorganisms are at equilibrium, and little net biomass growth occurs. In other words, biomass decay approximately balances biomass growth. Consequently, where no net biomass is produced, a stoichiometric equation describing microbial degradation of n-hexane is as follows:

$$C_6H_{14} + 9.5O_2 \rightarrow 6CO_2 + 7H_2O \qquad (3\text{-}6)$$

Based on Equation 3-6, 9.5 moles of oxygen are required for every mole of hydrocarbon consumed, or, on a weight basis, approximately 3.5 g of oxygen are required for every 1 g of hydrocarbon consumed.

To predict the amount of time required to bioremediate a site, it is necessary to understand the microbial kinetics of substrate (contaminant) utilization. Most substrate utilization falls under the heading of primary substrate utilization, in which growth on a carbon source supplies most of the carbon and energy for the microorganism. In cases where a contaminant does not supply the primary source or cannot be used for carbon and energy, secondary substrate utilization or cometabolism may occur. During the bioventing process, primary substrate utilization generally describes the kinetics of the reactions taking place; however, in some instances, cometabolic processes also may occur. For example, at sites contaminated with both fuels and solvents such as TCE, cometabolic bioventing may account for degradation of TCE.

Primary substrate utilization has been described through an empirical approach by the Monod expression:

$$-\frac{dS}{dt} = \frac{kXS}{K_s + S} \qquad (3\text{-}7)$$

where: S = concentration of the primary substrate (contaminant) (g_S/L);

t = time (minutes);

k = maximum rate of substrate utilization $(g_S/g_X\text{-min})$;

X = concentration of microorganisms (g_X/L); and

K_S = Monod half-velocity constant (g_S/L).

At high substrate concentrations $(S >> K_S)$, the rate of substrate utilization is at a maximum, limited by some other factor such as oxygen, nutrients, or the characteristics of the microorganism. In this instance, the rate of substrate utilization will be first-order with respect to cell density, but zero-order with respect to substrate concentration. Conversely, when the primary substrate concentration is very low $(S << K_S)$, the substrate utilization rate will be first-order with respect to both cell density and substrate concentration. In a well-designed bioventing system, kinetics based on oxygen utilization are zero-order. However, the rate based on petroleum or other contaminant removal may be described by Monod or inhibition kinetics.

Monod kinetics have been widely applied to conventional wastewater treatment where the compounds being treated generally are bioavailable and readily degradable. Bioventing typically is applied to aerobically biodegradable compounds; however, the maximum rate of biodegradation (k) is much lower than for most wastes in conventional wastewater treatment. For example, Howard et al. (1991) estimated that benzene has an aerobic half-life (dissolved in groundwater) of 10 days to 24 months, whereas ethanol (a compound more typical of conventional wastewater treatment) is estimated to have a half-life of 0.5 to 2.2 days. Bioventing kinetics are further complicated by bioavailability of the contaminants, driven at least in part by solubilization. Since microorganisms exist in pore water, contaminants must partition into the pore water to be available to be degraded. Although high soil contaminant concentrations may be present, the actual concentration of hydrocarbon dissolved in the pore water and available to the microorganisms may be low.

In practice, oxygen utilization rates tend to decline slowly with time during remediation. At many sites, this trend may be difficult to follow over periods of less than 1 to 3 years because of other variables affecting the rate, such as temperature and soil moisture. This decline may not be indicative of true first-order kinetics, but simply may be the result of selective early removal of more degradable compounds such as benzene.

B. ENVIRONMENTAL PARAMETERS AFFECTING MICROBIAL PROCESSES

The success of a bioventing system is dependent upon providing microorganisms optimal conditions for active growth. Several factors may affect a microorganism's ability to degrade contaminants, including:

- availability and type of electron acceptors;
- moisture content;
- soil pH;
- soil temperature;
- nutrient availability; and
- contaminant concentration.

Each of these parameters was measured at U.S. Air Force Bioventing Initiative sites. The actual effect of individual parameters on microbial activity is difficult to assess in the field due to interference and interactions among these parameters. The in situ respiration test was used as a measurement tool that integrates all factors to assess whether the microorganisms are metabolizing the fuel. Data from the in situ respiration test and site measurements were used to conduct a statistical analyses of the observed effects of the site measurements on microbial activity in the field. The statistical analysis was constructed to account for parameter interactions. These results are discussed in detail in Appendix A. A more general discussion of the significance of each of these parameters and its effect on microbial activity is provided in the following sections.

1. Electron Acceptor Conditions

One of the most important factors which influences the biodegradability of a compound is the type and availability of electron acceptors. For example, following a hydrocarbon spill, as a result of the hydrocarbon biodegradability, anaerobic conditions typically predominate in the subsurface because of oxygen depletion from microbial activity. While hydrocarbons may undergo limited biodegradation under anaerobic conditions (Bilbo et al., 1992; Mormile et al., 1994), in general, aerobic conditions are most suitable for relatively rapid remediation of petroleum hydrocarbons. Therefore, oxygen supply is critical to the success of a bioventing system. In field studies, oxygen has been found to be the most important factor in determining the success of a bioventing system (Hinchee et al., 1989b; Miller et al., 1991). This has been confirmed by the data collected as part of the U.S. Air Force Bioventing Initiative. At these sites, oxygen has been found to the primary factor limiting microbial activity at all but three sites (Miller et al., 1993).

2. Moisture Content

Soil moisture content may impact the bioventing process by its effect on microorganisms or on soil gas permeability. Microorganisms require moisture for metabolic processes and for solubilization of energy and nutrient supplies. Conversely, soil moisture content directly affects the soil gas permeability, with high moisture contents resulting in poor distribution of oxygen. In practice, low soil moisture content has been found to directly limit biodegradation rates only where bioventing has been implemented in very dry desert environments. A more common influence of moisture is that excess moisture has led to significant reductions in soil gas permeability. One of the major objectives of the U.S. Air Force Bioventing Initiative was to assess the effects of moisture on biodegradation.

The range of soil moisture content measured at U.S. Air Force Bioventing Initiative sites is shown in Figure 3-4. The lowest soil moisture content measured was 2% by weight, and microbial activity still was observed in these soils. Figure 3-5 illustrates the observed relationship between soil moisture and oxygen utilization rates. To date, a strong correlation has not been recorded between moisture content and oxygen utilization rate, although a slight positive relationship has been observed as discussed in Appendix A.

At a desert site at the Marine Corps Air Ground Combat Center, Twentynine Palms, California, soil moisture content appeared to detrimentally affect microbial activity. Soil moisture content ranged from 2% to 4% by weight and, although the site was contaminated with jet fuel, significant oxygen limitation was not observed. An irrigation system was installed at the site in an effort to enhance microbial activity. The site was irrigated for 1 week, then bioventing was initiated for 1 month before conducting an in situ respiration test. In situ respiration rates measured after irrigation were significantly higher than those measured prior to irrigation (Figure 3-6). In addition, prior to irrigation, oxygen utilization did not occur once oxygen concentrations dropped below approximately 17%. After irrigation, activity continued until oxygen was completely consumed to less than 1%. These results demonstrated that, in extreme cases, moisture addition may improve the performance of bioventing systems through enhanced microbial activity.

A frequent concern pertaining to soil moisture content is that air injection bioventing will dry out the soil to a point what would be detrimental to microbial growth, necessitating humidification of the injection air. However, a simple calculation as shown in Example 3-1 illustrates that over a 3-year period, moisture loss is minimal. Drying and

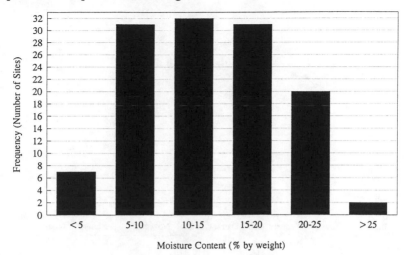

Figure 3-4. Soil Moisture Content Measurements at Bioventing Initiative Sites

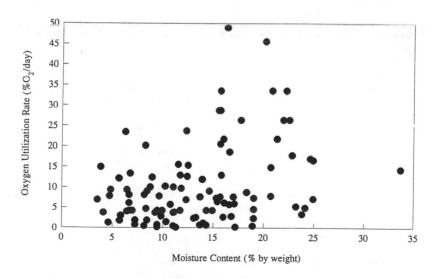

**Figure 3-5.   Direct Correlation Between Oxygen Utilization Rates and Soil Moisture
Content at Bioventing Initiative Sites**

30

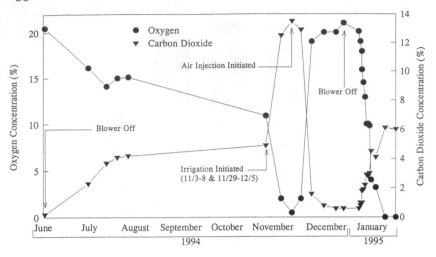

Figure 3-6. Oxygen and Carbon Dioxide Concentrations Prior to and After Irrigation at Twentynine Palms, California

moisture loss as a result of bioventing usually are a problem only very near the vent well or if very high air injection rates are used (typically not the case in properly designed bioventing systems). Sites typically have several moisture sources that also make drying due to air injection negligible, such as rain and snow, and water as a byproduct of mineralization (generated at a rate of 1.5 kg water for every 1 kg of hydrocarbon degraded as shown in Equation 3-6).

Example 3-1. Moisture Content Change During Air Injection and Water Generated During Biodegradation: For this test:

Vapor pressure (P_{water})	=	17.5 mm Hg
Flowrate (Q)	=	1 pore volume/day, typical of bioventing
Volume of treatment area (V)	=	12,300 m^3
Biodegradation rate (k_B)	=	3 mg/kg-day
Initial moisture content	=	15% by weight
Soil bulk density	=	1,440 kg/m^3

Assume worst case of 0% humidity and no infiltration.

To calculate the total water at the site initially, the mass of soil is first calculated:

$$12,300 \text{ m}^3 \times \frac{1,440 \text{ kg}}{\text{m}^3} = 1.8 \times 10^7 \text{ kg soil}$$

Therefore, the initial mass of water is:

$$(1.8 \times 10^7 \text{ kg soil}) \times 0.15 = 2.7 \times 10^6 \text{ kg } H_2O$$

Since the flowrate is equivalent to one pore volume/day, the mass of water removed per day will be equal to the mass of water in the vapor phase of the treated area, which can be calculated using the Ideal Gas law:

$$\frac{\text{Moles } H_2O \text{ removed}}{\text{day}} = \frac{P_{water} \ V}{RT} = \frac{17.5 \text{ mmHg} \times 12,300 \text{ m}^3}{\left(0.0623 \ \dfrac{m^3 - mmHg}{mole - ^\circ K}\right) \times 298^\circ K}$$

$$\frac{\text{Moles } H_2O \text{ removed}}{\text{day}} = 11,600 = 210 \ \frac{\text{kg}}{\text{day}}$$

Total water removal in 3 years:

$$210 \ \frac{\text{kg}}{\text{day}} \times 1,095 \text{ days} = 230,000 \text{ kg removed}$$

This water loss represents a fairly small percentage, or:

$$\frac{230,000 \text{ kg evaporated}}{2.7 \times 10^6 \text{ initial mass}} = 0.086 \approx 8.6\% \ H_2O \text{ loss}$$

This is equivalent to a soil moisture drop from approximately 15% to 13.7%. Assuming a contaminated thickness of 10 ft (3 m), an infiltration rate of approximately 2.4 inches (6.1 cm) in 3 years, or less than 1 inch (2.5 cm) per year, would replace the lost moisture. In practice, some drying very close to the vent well may be observed, but usually is not.

Water loss also will be replaced through biodegradation of hydrocarbons. Calculating the total mass of hydrocarbons degraded over 3 years:

$$3 \ \frac{mg}{kg - day} \times 1,095 \text{ days} \times (1.8 \times 10^7 kg \text{ soil}) \times \frac{kg}{10^6 mg} = 59,000 \text{ kg hydrocarbon degraded}$$

Based on the stoichiometry in Equation 3-6 where 1.5 kg of water are generated for every kg of hydrocarbon degraded, the amount of water generated would be:

$$59,000 \text{ kg hydrocarbon} \times \frac{1.5 \text{ kg water}}{\text{kg hydrocarbon}} = 88,500 \text{ kg water}$$

Therefore, total water removal in three years must also account for the water generation, where:

$$230,000 \text{ kg} - 88,500 \text{ kg } H_2O = 141,500 \text{ kg } H_2O \text{ loss}$$

This is equivalent to a water loss of 5.3% over 3 years.

3. Soil pH

Soil pH also may affect the bioremediation process, since microorganisms require a specific pH range in order to survive. Most bacteria function best in a pH range between 5 and 9 with the optimum being slightly above 7 (Dragun, 1988). A shift in pH may result in a shift in the makeup of the microbial population, because each species will exhibit optimal growth at a specific pH. Throughout the U.S. Air Force Bioventing Initiative, pH has not been found to limit in situ bioremediation, and is probably only of concern where contamination has radically altered the existing pH.

Figure 3-7 illustrates the range of soil pH found at the U.S. Air Force Bioventing Initiative sites to date. In general, the majority of sites have fallen within the "optimal" pH range for microbial activity of 5 to 9. However, microbial respiration based on oxygen utilization has been observed at all sites, even in soils where the pH was below 5 or above 9. Figure 3-8 illustrates the observed relationship between pH and oxygen utilization rates. Based upon these observations, it appears that pH is not a concern when bioventing at most sites.

4. Soil Temperature

Microbial activity has been reported at temperatures varying from -12 to 100°C (10 to 212°F) (Brock et al., 1984); however, the optimal range for biodegradation of most contaminants is generally much narrower and will vary by species. For example, microorganisms in a subarctic environment may exhibit optimal growth at 10°C (50°F), whereas microorganisms in a subtropical environment may exhibit optimal growth at 30°C (86°F).

It has generally been observed that biodegradation rates double for every 10°C (50°F) temperature increase, up to some inhibitory temperature. The van't Hoff-Arrhenius equation expresses this relationship quantitatively as:

$$k_T = k_o\ e^{\frac{-E_a}{RT_{abs}}} \tag{3-8}$$

where:
k_T = temperature-corrected biodegradation rate (% O_2/day)
k_o = baseline biodegradation rate (% O_2/day)
E_a = activation energy (cal/mole)
R = gas constant (1.987 cal/°K-mol)
T_{abs} = absolute temperature (°K)

Miller (1990) found E_a equal to 8 to 13 kcal/mole for in situ biodegradation of jet fuel. In the 17 to 27°C (63 to 81°F) range, the van't Hoff-Arrhenius relationship accurately predicted biodegradation rates. A similar analysis was conducted at Site 20, Eielson AFB, Alaska, where the activation energy was found to be equal to 13.4 kcal/mole (Example 3-2).

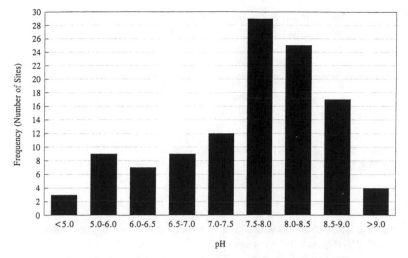

Figure 3-7. Soil pH Measurements at Bioventing Initiative Sites

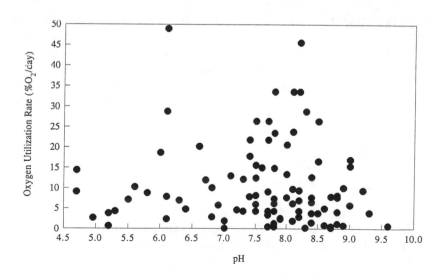

Figure 3-8.   Correlation Between Oxygen Utilization Rate and Soil pH at Bioventing
             Initiative Sites

Figure 3-9 illustrates the relationship between oxygen utilization rate and temperature observed at Site 20, Eielson AFB, a JP-4 jet fuel-contaminated site.

Example 3-2.   Calculation of the van't Hoff-Arrhenius Constant From Site Data:
Various forms of soil warming were tested at Site 20, Eielson AFB, Alaska.   This resulted in a wide range of temperatures and biodegradation rates measured at the same site.

In order to calculate the van't Hoff-Arrhenius constant, the log of the biodegradation rate must be calculated versus the inverse of the temperature to provide the relationship:

$$\ln\left(\frac{k_T}{k_o}\right) = \frac{-E_a}{R} \times \frac{1}{T} \tag{3-9}$$

The slope of the linear regression of inverse temperature versus oxygen utilization rate is $-6740\,°K$ (Figure 3-9).   Therefore,

$$\frac{E_a}{R} = -6740$$

$$\frac{E_a}{1.987\dfrac{cal}{°K-mole}} = -6740$$

$$E_a = 1.987\frac{cal}{°K-mole} \times -6740 = 13,390\frac{cal}{mole} \approx 13.4\frac{kcal}{mole}$$

Heat addition may improve bioventing performance by increasing microbial activity.   Solar warming, warm water infiltration, and buried heat tape have been used to increase soil temperature.   Their use has resulted in increased microbial activity and contaminant degradation (Leeson et al., 1995).   Selection of a soil warming technique will depend on cost considerations versus remediation time requirements[1].   While warm water infiltration or heat tape can significantly increase biodegradation rates, the cost is significantly higher than simply using surface insulation or no heating method.   The use of warm water infiltration, although effective, is limited to very permeable soils, to ensure that adequate drainage of the applied water will occur.   The use of soil heating to increase biodegradation rates may prove cost effective only in cold regions, such as Alaska.

5. Nutrient Supply
In order to sustain microbial growth, certain nutrients must be available at minimum levels.   The following nutrients/cofactors are known to be

---

[1]   Refer to Chapter 4 for a discussion of the cost benefit of soil warming.

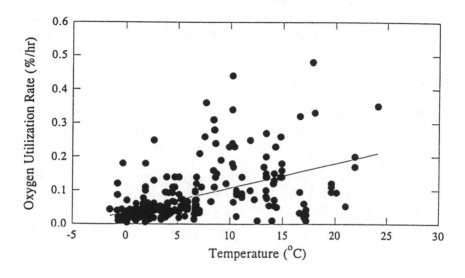

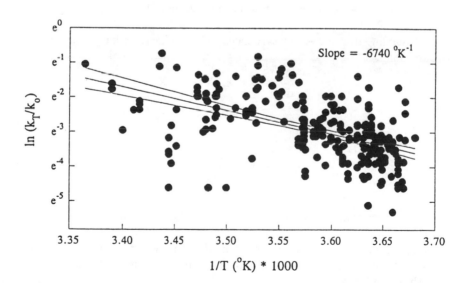

Figure 3-9. **Soil Temperature Versus Biodegradation Rate at Site 20, Eielson AFB, Alaska**

required in order to support microbial growth: calcium, cobalt, copper, iron, magnesium, manganese, molybdenum, nitrogen, phosphorus, potassium, sodium, sulfur, and zinc. Nitrogen and phosphorus are required in the greatest concentrations and are the nutrients most likely to be limiting. The remaining chemicals are considered micronutrients, because they are required in only small quantities and generally are available in excess in nature.

Nutrients are required as components of the microbial biomass. The need for these nutrients is very different from the need for oxygen (or other electron donors) and the carbon source. Nutrients are not destroyed, but are recycled by the ecosystem. Thus, unlike oxygen, a steady input of nutrients is not required.

An approach to estimation of nutrient requirements, suggested by John T. Wilson of the U.S. EPA Ada Laboratory, can be made based on microbial kinetics. Starting with:

$$\frac{dX}{dt} = k_B Y - k_d X \qquad (3-10)$$

where: X = biomass (mg biomass/kg soil)
      k_B = biodegradation rate (mg hydrocarbon/kg soil-day)
      Y = cell yield (mg biomass/mg hydrocarbon)
      k_d = endogenous respiration rate (day^{-1})

Assuming that the biomass concentration achieves steady state during bioventing,

$$\frac{dX}{dt} = 0 = k_B Y - k_d X \qquad (3-11)$$

Solving:

$$X = \frac{k_B Y}{k_d} \qquad (3-12)$$

Little is known about the in situ cell yields or endogenous respiration rates of hydrocarbon-degrading organisms, but these parameters can be estimated based on ranges reported in the wastewater treatment literature (Metcalf and Eddy, 1979). An example for calculating required nutrients is shown in Example 3-3.

Example 3-3. Estimation of Nutrient Requirements In Situ: For a given site, the following is assumed:

k_B = 10 mg/kg-day (typical rate found at bioventing sites)
Y = 0.5 mg/mg
k_d = 0.05/day

Solving:

$$X = \frac{10 \; \frac{mg}{kg-day} \times 0.5 \; \frac{mg}{mg}}{\frac{0.05}{day}} = 100 \; \frac{mg}{kg}$$

To sustain 100 mg/kg of biomass, the nutrient requirements may be estimated from biomass to nutrient ratios. A variety of ratios are found in the literature. For this example, a 100:10:1 ratio of biomass:nitrogen:phosphorus is assumed. This ratio yields a nutrient requirement of 10 mg/kg of nitrogen and 1 mg/kg of phosphorus. Thus, if the above assumptions hold, a site with at least these levels of nitrogen and phosphorus initially should not be rate-limited by nitrogen and phosphorus.

Most soils naturally contain nutrients in excess of the concentrations calculated in Example 3-3. Therefore, although the addition of nutrients may be desirable in hopes of increasing biodegradation rates, field research to date does not indicate the need for these additions (Dupont et al., 1991; Miller et al., 1991). Nutrients are often added to bioremediation projects in anticipation of increased biodegradation rates; however, field data to date do not show a clear relationship between increased rates and supplied nutrients.

Concentrations of total Kjeldahl nitrogen (TKN) and total phosphorus at U.S. Air Force Bioventing Initiative sites and the corresponding relationship between oxygen utilization rates are shown in Figures 3-10 through 3-13. Although optimal ratios of carbon, nitrogen, and phosphorus were not available at all sites, the natural nutrient levels were sufficient to sustain some level of biological respiration at all sites when the most limiting element, oxygen, was provided.

In controlled nutrient additions at Tyndall and Hill AFBs, no apparent increase in microbial activity was observed. Therefore, there appeared to be no benefit of nutrient addition. The relationship between oxygen utilization rates and TKN and total phosphorus are shown in Figures 3-11 and 3-13, respectively. As is illustrated in these figures, there is no correlation between phosphorus and oxygen utilization rates and only a weak relationship between TKN concentrations and oxygen utilization rates, again emphasizing that natural ambient nutrient levels seem sufficient for microbial activity.

Figure 3-14 illustrates the range of iron concentrations measured at U.S. Air Force Bioventing Initiative sites. Iron concentrations varied greatly, with concentrations from less than 100 mg/kg to greater than 75,000

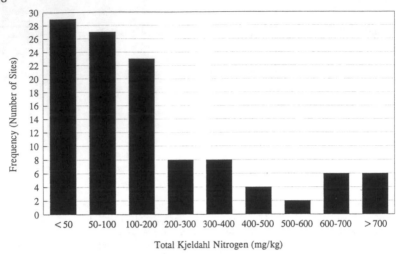

Figure 3-10. TKN Measurements at U.S. Air Force Bioventing Initiative Sites

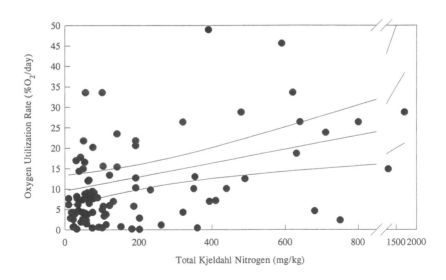

Figure 3-11. Correlation Between Oxygen Utilization Rate and TKN at U.S. Air Force Bioventing Initiative Sites

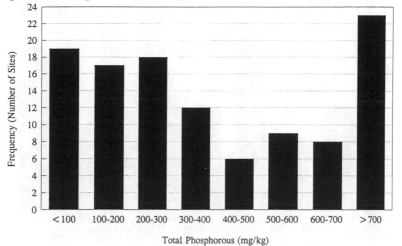

Figure 3-12. Total Phosphorus Measurements at U.S. Air Force Bioventing Initiative Sites

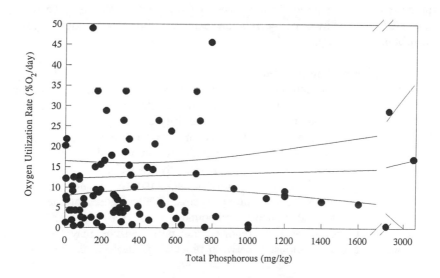

Figure 3-13. Correlation Between Oxygen Utilization Rate and Total Phosphorus at U.S. Air Force Bioventing Initiative Sites

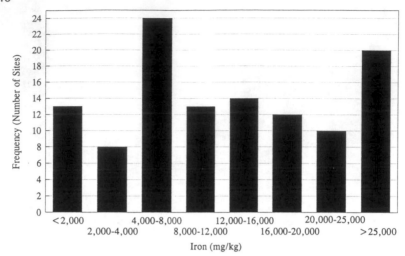

Figure 3-14. Iron Concentration Measurements at U.S. Air Force Bioventing Initiative Sites

mg/kg. Soils in Hawaii and Alaska exhibited the highest iron contents. Although iron is a nutrient required for microbial growth, iron also may react with oxygen to form iron oxides. Theoretically, if a significant amount of iron oxidation occurs, the observed oxygen utilization rate[1] would not reflect microbial activity only. Therefore, calculated biodegradation rates would be an overestimate of actual biodegradation rates. Thus, background wells in uncontaminated areas are recommended in bioventing applications in areas of high iron concentrations. To date, this study has shown no correlation between iron content and oxygen utilization rates (Figure 3-15).

6. Contaminant Concentration

Contaminant concentration also may affect biodegradation of the contaminant itself. Excessive quantities of a contaminant can result in a reduction in biodegradation rate due to a toxicity effect. Conversely, very low concentrations of a contaminant also may reduce overall degradation rates because contact between the contaminant and the microorganism is limited and the substrate concentration is likely below S_{min}.

In practice, petroleum hydrocarbons in fuel mixtures do not generally appear to be toxic to the bioventing process. Other more soluble (i.e., phenolics) or less biodegradable compounds (i.e., TCE) may exhibit a toxicity effect and it has been reported that pure benzene may be toxic. Although a general relationship between bioventing rates and concentration

---

[1] As measured by in situ soil gas oxygen concentrations.

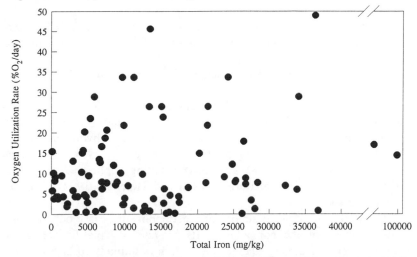

Figure 3-15. Correlation Between Oxygen Utilization Rates and Iron Content at U.S. Air Force Bioventing Initiative Sites

no doubt exists, the relationship is complex and not fully understood. At sites where NAPLs are present (soil concentrations above the 100 to 1,000 mg/kg range), the bioavailable hydrocarbon is most probably limited by solubilization, which is linked to Raoult's law and, to an extent, is independent of total hydrocarbon concentration. Certainly, the NAPL distribution can affect the proportion of the soil in a site in which biodegradation is occurring, and at lower concentrations less soil may be in direct contact with NAPLs. It is likely that the reduction in biodegradation rates observed over time on many sites is due at least in part to changes in the hydrocarbon makeup as the more degradable and more mobile compounds (i.e., benzene, toluene, ethylbenzene, and xylenes) are removed. At lower hydrocarbon concentrations where NAPLs are not present, a decline in rate would be expected with time as the available substrate is removed.

7. Bioavailability and Relative Biodegradability

Another critical parameter affecting the extent of in situ bioremediation is bioavailability of the contaminant(s) of concern. Bioavailability is a general term to describe the accessibility of contaminants to the degrading populations. Bioavailability consists of (1) a physical aspect related to phase distribution and mass transfer, and (2) a physiological aspect related to the suitability of the contaminant as a substrate (U.S. EPA, 1993, EPA/540/S-93/501). Compounds with greater aqueous solubilities and lower affinities to partition into NAPL or to sorb onto the soil generally are bioavailable to soil microorganisms and are more readily degraded. For example, BTEX is preferentially degraded relative to the larger alkanes found in fuels. The most likely explanation for this is that BTEX is more mobile and more soluble in pore water and therefore is more bioavailable.

III. COMPOUNDS TARGETED FOR REMOVAL

Any aerobically biodegradable compound, such as petroleum hydrocarbons, potentially can be degraded though bioventing. To date, bioventing has been applied primarily to petroleum hydrocarbons (see Table 2-2); however, bioventing of PAHs (Lund et al., 1991; Hinchee and Ong, 1992; Alleman et al., 1995) and bioventing applied to an acetone, toluene, and naphthalene mixture (Leeson and Hinchee, 1994) have been implemented successfully.

The key to bioventing feasibility in most applications is biodegradability versus volatility of the compound. If the rate of volatilization greatly exceeds the rate of biodegradation, bioventing is unlikely to be successful, as removal occurs primarily through volatilization. This will occur most often in those cases where the contaminant is a fresh, highly volatile fuel. An unsuccessful bioventing application is unlikely to occur due to a lack of microbial activity. If bioventing is operated in the injection mode as recommended by this document, volatilized contaminants may be biodegraded before reaching the surface, unlike during an extraction operation. Figure 3-16 illustrates the relationship between a compound's physicochemical properties and its potential for bioventing.

In general, compounds with a low vapor pressure[1] cannot be successfully removed by volatilization, but can be biodegraded in a bioventing application if they are aerobically biodegradable. High vapor pressure compounds are gases at ambient temperatures. These compounds volatilize too rapidly to be easily biodegraded in a bioventing system, but typically are a small component of fuels and, due to their high volatility, will attenuate rapidly. Compounds with vapor pressures between 1 and 760 mmHg may be amenable to either volatilization or biodegradation. Within this intermediate range lie many of the petroleum hydrocarbon compounds of greatest regulatory interest, such as benzene, toluene, ethylbenzene, and the xylenes. As can be seen in Figure 3-16, various petroleum fuels are more or less amenable to bioventing. Some components of gasoline are too volatile to easily biodegrade, but, as stated previously, typically are present in low overall concentrations and are attenuated rapidly. Most of the diesel constituents are sufficiently

---

[1] For the purposes of this discussion, compounds with vapor pressures below approximately 1 mmHg are considered low, and compounds with vapor pressures above approximately 760 mmHg are considered high.

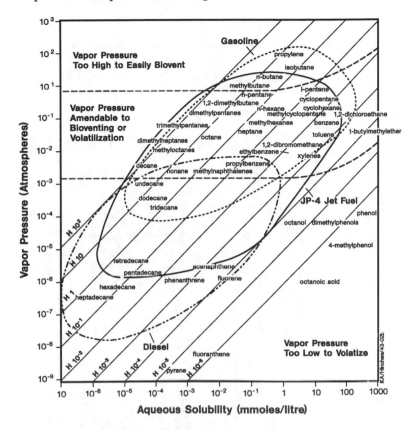

Figure 3-16. Relationship Between Contaminant Physicochemical Properties and Potential for Bioventing

H - Henry's Law Coefficient (atm m³/mole)

nonvolatile to preclude volatilization, whereas the constituents of JP-4 jet fuel are intermediate in volatility.

To be amenable to bioventing, a compound must (1) biodegrade aerobically at a rate resulting in an oxygen demand greater than the rate of oxygen diffusion from the atmosphere, and (2) biodegrade at a sufficiently high rate to allow in situ biodegradation before volatilization. Practically, this means that low vapor pressure compounds need not biodegrade as rapidly as high vapor pressure compounds for bioventing to be successful. Figure 3-17 illustrates this relationship. The actual feasibility of bioventing is very site-specific and Figures 3-16 and 3-17 should not be used as absolutes, but rather as general guidelines.

Bioventing generally is not considered appropriate for treating compounds such as polychlorinated biphenyls (PCBs) and chlorinated hydrocarbons. However, through a cometabolic process, it may be

44

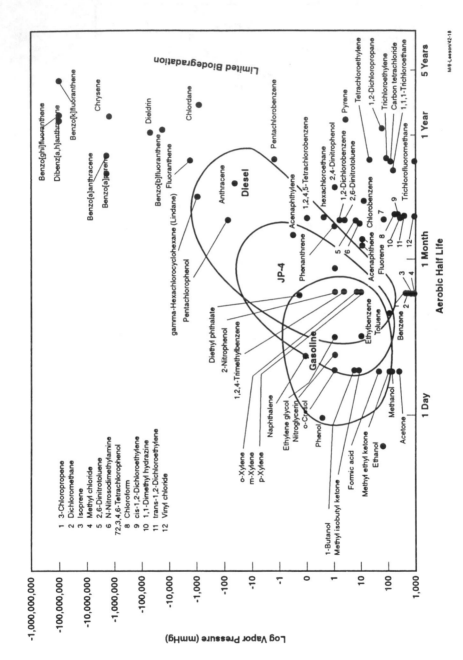

Figure 3-17. Relationship Between Contaminant Vapor Pressure and Aerobic Biodegradability

possible to enhance the degradation of compounds such as TCE through bioventing. In laboratory studies, it has been shown that if toluene is present to provide the primary source of carbon, organisms that grow on toluene may be able to cometabolize TCE (Wackett and Gibson, 1992). More recently, Hopkins et al. (1993) demonstrated TCE degradation in situ through the injection of oxygen and phenol into an aquifer. TCE removal of 88% was observed in the field, indicating the potential for cometabolic degradation of chlorinated compounds in situ.

IV. BTEX VERSUS TPH REMOVAL IN PETROLEUM-CONTAMINATED SITES

BTEX generally are the compounds that are regulated most stringently. Typically, these compounds degrade very rapidly during bioventing, and at most sites, are degraded to below detection limits within 1 year of operation of a bioventing system. This trend was illustrated in the study at Tyndall AFB described in Chapter 4 and has been confirmed at 81 sites completing the 1-year testing under the U.S. Air Force Bioventing Initiative. At Tyndall AFB, two test plots were studied with initial hydrocarbon concentrations of 5,100 and 7,700 mg/kg. After 9 months of bioventing, TPH was reduced by 40% from the initial concentration. However, the low-molecular-weight compounds such as BTEX were reduced by more than 90% (Figure 3-18). The low-molecular-weight compounds were preferentially degraded over the heavier fuel components, which is consistent with previous research (Atlas, 1986).

If a risk-based approach to remediation is used that focuses on removing the soluble, mobile, and more toxic BTEX components of the fuel, remediation times can be reduced significantly, making bioventing an attractive technology for risk-based remediations. In addition, the BTEX compounds often are initially at relatively low levels at many fuel-contaminated sites as illustrated by results from the U.S. Air Force Bioventing Initiative. Data collected from the majority of the U.S. Air Force Bioventing Initiative sites demonstrate that more than 85% of initial soil samples contained less than 1 mg/kg of benzene (Figure 3-19). An exception to this may be gasoline-contaminated sites; the majority of sites included in the U.S. Air Force Bioventing Initiative were contaminated with heavier weight contaminants. Only 19 of 125 U.S. Air Force Bioventing Initiative sites were contaminated by gasoline or AVGAS.

46

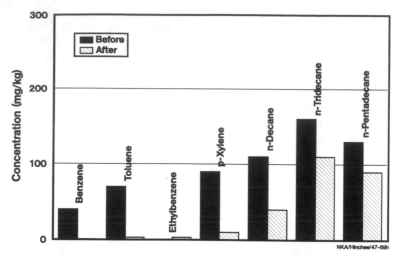

Figure 3-18. Results of Soil Analysis Before and After Venting from Plot V2 at Tyndall AFB, Florida

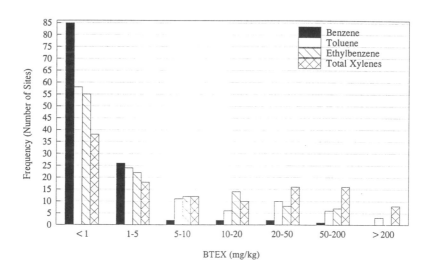

Figure 3-19. Contaminant Distribution at Bioventing Initiative Sites

CHAPTER 4

CASE HISTORIES OF EARLY BIOVENTING STUDIES

Four of the first well-documented bioventing studies are presented in this section to illustrate significant results which have contributed to the development of bioventing, the U.S. Air Force Bioventing Initiative, and this document. The development of the U.S. Air Force Bioventing Initiative was based largely upon the results from these four early studies, which are discussed in detail in this section. Site 914, Hill AFB, Utah, was one of the first bioventing systems studied. This study was designed to examine the feasibility of biodegradation through air injection, and also to investigate the effect of nutrient and moisture addition on biodegradation. The second site was a bioventing system at Tyndall AFB, Florida, initiated in 1990. This study was short-term (9 months), but was designed to examine process variables in more detail than was possible at Site 914, Hill AFB. The third site discussed in this chapter was conducted at Site 280, Hill AFB, Utah. This study was initiated in 1991 as a bioventing site and was operated for approximately 3 years. Research on air flowrates and injection depth was carried out at this site. The fourth study presented in this chapter was conducted at Eielson AFB, Alaska. This study was initiated in 1991 as a bioventing system and was operated for 3 years. This study was conducted to examine the feasibility of bioventing in a subarctic climate as well as to evaluate the effects of soil warming on biodegradation rates.

These case histories are not presented as design examples, since they were designed as research efforts. In fact, these studies have been the basis for development of current design practice as presented in this document. Details of each of these studies are presented in the following sections.

I. SITE 914, HILL AFB, UTAH

A spill of approximately 27,000 gallons of JP-4 jet fuel occurred at Site 914 when an automatic overflow device failed. Contamination was limited to the upper 65 ft (20 m) of a delta outwash of the Weber River. This surficial formation extends from the surface to a depth of approximately 65 ft (20 m) and is composed of mixed sand and gravel with occasional clay stringers. Depth to regional groundwater is approximately 600 ft (180 m);

47

however, water occasionally may be found in discontinuous perched zones. Soil moisture averaged less than 6% by weight in the contaminated soils.

The collected soil samples had JP-4 jet fuel concentrations up to 20,000 mg/kg, with an average concentration of approximately 400 mg/kg (Oak Ridge National Laboratory, 1989). Contaminants were unevenly distributed to depths of 65 ft (20 m). Vent wells were drilled to approximately 65 ft (20 m) below the ground surface and were screened from 10 to 60 ft (3 to 20 m) below the surface. A background vent well was installed in an uncontaminated location in the same geologic formation approximately 700 ft (210 m) north of the site.

This system originally was designed for SVE, not for bioventing. During the initial 9 months of operation, it was operated to optimize volatilization, while biodegradation was merely observed. After this period, air flowrates were greatly reduced, and an effort was made to optimize biodegradation and limit volatilization.

Soil vapor extraction was initiated in December 1988 at a rate of approximately 25 cubic ft per minute (cfm) (710 L/min). The off-gas was treated by catalytic incineration, and initially it was necessary to dilute the highly concentrated gas to remain below explosive limits and within the incinerator's hydrocarbon operating limits. The venting rate was gradually increased to approximately 1,500 cfm (4.2×10^4 L/min) as hydrocarbon concentrations dropped. During the period between December 1988 and November 1989, more than 3.5×10^8 ft^3 (9.9×10^{10} L) of soil gas were extracted from the site.

In November 1989, ventilation rates were reduced to between approximately 300 and 600 cfm (8,500 to 17,000 L/min) to provide aeration for bioremediation while reducing off-gas generation. This change allowed removal of the catalytic incinerators, saving approximately $13,000 per month in rental and propane costs.

Hinchee and Arthur (1991) conducted bench-scale studies using soils from this site and found that, in the laboratory, both moisture and nutrients appeared to become limiting after aerobic conditions had been achieved. These findings led to the addition of first moisture and then nutrients in the field. Moisture addition clearly stimulated biodegradation, whereas nutrient addition did not (Figure 4-1). The failure to observe an effect of nutrient addition could be explained by a number of factors:

1. The nutrients failed to move in the soils, which is a problem particularly for ammonia and phosphorus (Aggarwal et al., 1991).
2. Remediation of the site was entering its final phase and nutrient addition may have been too late to result in an observed change.
3. Nutrients simply may have not been limiting.

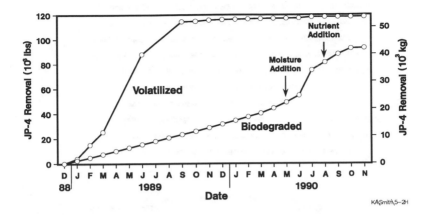

<figure>

Figure 4-1. Cumulative Hydrocarbon Removal and the Effect of Moisture and Nutrient Addition at Site 914, Hill AFB, Utah
</figure>

During extraction, oxygen and hydrocarbon concentrations in the off-gas were measured. To quantify the extent of biodegradation at the site, the oxygen was converted to an equivalent basis. This was based on the stoichiometric oxygen requirement for hexane mineralization[1]. Hydrocarbon concentrations were determined based on direct readings of a total hydrocarbon analyzer calibrated to hexane. Based on these calculations, the mass of JP-4 jet fuel as carbon removed was approximately 1,500 pounds volatilized and 93,000 pounds biodegraded (Figure 4-1). After a 2-year period, cleanup and regulatory closure were achieved (Figure 4-2).

The results of this study indicated that aerobic biodegradation of JP-4 jet fuel did occur in the vadose zone at Site 914. Biodegradation was increased by soil venting at this site because, prior to venting, biodegradation appeared to have been oxygen limited. The SVE system, designed to volatilize the fuel, stimulated in situ biodegradation with no added nutrients or moisture. In this study, approximately 15% of the documented field removal observed at the site was the result of microbial-mediated mineralization to carbon dioxide. Additional biological fuel removal by conversion to biomass and degradation products no doubt occurred, but was not quantified.

From this study, it was apparent that further studies of field biodegradation in unsaturated soils were needed to develop a better understanding of the effects of such variables as oxygen content, nutrient

---

[1] Refer to Chapter 6 for a discussion of this calculation.

50

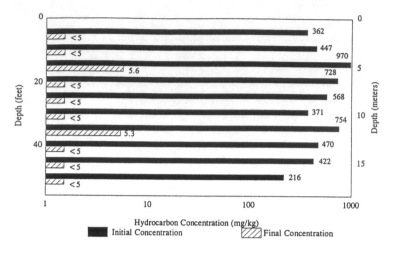

Figure 4-2. Results of Soil Analysis Before and After Treatment at Site 914, Hill AFB, Utah

requirements, soil moisture, contaminant levels, and soil type on the limitation and optimization of bioventing of contaminated field sites. Also, further studies of gas transport in the vadose zone were needed to ensure adequate design of air delivery systems.

Further details of this study may be found in Dupont et al. (1991) and Hinchee et al. (1991b).

II. TYNDALL AFB, FLORIDA

A more controlled study than was possible at Site 914, Hill AFB was designed at Tyndall AFB as a follow-up to the Hill AFB research. The experimental area in the Tyndall AFB study was located at a site where past JP-4 jet fuel storage had resulted in contaminated soils. The nature and volume of fuel spilled or leaked were unknown. The site soils were a fine- to medium-grained quartz sand. The depth to groundwater was 2 to 4 ft (0.61 to 1.2 m).

The field study was designed with the following objectives:

- to determine whether bioventing enhanced biodegradation of JP-4 jet fuel at this site;
- to determine whether moisture addition coupled with bioventing enhanced biodegradation rates;
- to determine whether nutrient addition coupled with bioventing enhanced biodegradation rates;
- to evaluate flowrate manipulation to maximize biodegradation and minimize volatilization; and

- to calculate specific biodegradation rate constants from a series of respiration tests conducted during shutdown of the air extraction system.

Four test cells were constructed to allow control of gas flow, water flow, and nutrient addition. Test cells V1 and V2 were installed in the hydrocarbon-contaminated zone; test cells V3 and V4 were installed in uncontaminated soils. Test cells were constructed and operated in the following manner:

- V1 (uncontaminated): Venting for approximately 8 weeks, followed by moisture addition for approximately 14 weeks, followed by moisture and nutrient addition for approximately 7 weeks.
- V2 (uncontaminated): Venting coupled with moisture and nutrient addition for 29 weeks.
- V3 (uncontaminated): Venting with moisture and nutrient addition at rates similar to V2, with injection of hydrocarbon-contaminated off-gas from V1. Operation was conducted at a series of flowrates and retention times.
- V4 (uncontaminated): Venting with moisture and nutrient addition at rates similar to V2.

Initial site characterization indicated the mean soil hydrocarbon levels were 5,100 and 7,700 mg of hexane-equivalent/kg in treatment plots V1 and V2, respectively. The contaminated area was dewatered, and hydraulic control was maintained to keep the depth to water at approximately 5.25 ft (1.6 m). This exposed more of the contaminated soil to aeration. During normal operation, air flowrates were maintained at approximately one air-filled void volume per day.

Biodegradation and volatilization rates were much higher at the Tyndall AFB site than those observed at Hill AFB. These higher rates were likely due to higher average levels of contamination, higher temperatures, and higher moisture content. Biodegradation rates during bioventing ranged from approximately 2 to 20 mg/kg-day, with average values of 5 mg/kg-day. After 200 days of aeration, an average hydrocarbon reduction of approximately 2,900 mg/kg was observed. This represented a reduction in total hydrocarbons of approximately 40%.

Another important observation of this study was the effect of temperature on the biodegradation rate. Miller (1990) found that the van't Hoff-Arrhenius equation provided an excellent model of temperature effects. In the Tyndall AFB study, soil temperature varied by only approximately 7°C (44.6°F), yet biodegradation rates were approximately twice as high at 25°C (77°F) than at 18°C (64.4°F).

Operational data and biodegradation rates indicated that soil moisture and nutrients were not limiting factors in hydrocarbon biodegradation for this site (Figure 4-3). The lack of moisture effect contrasts with the Hill AFB findings, but most likely is the result of contrasting climatic and hydrogeologic conditions. Hill AFB is located in a high-elevation desert with a very deep water table. Tyndall AFB is located in a moist, subtropical environment, and at the site studies, the water table was maintained at a depth of approximately 5.25 ft (1.6 m). The nutrient findings support field observations at Hill AFB that the addition of nutrients does not stimulate biodegradation. Based on acetylene reduction studies, Miller (1990) speculated that adequate nitrogen was present due to nitrogen fixation. Both the Hill and Tyndall AFB sites had been contaminated for several years before the bioventing studies began, and both sites were anaerobic. It is possible that nitrogen fixation, which is maximized under these conditions, provided the required nutrients. In any case, these findings show that nutrient addition is not always required.

In the Tyndall AFB study, a careful evaluation of the relationship between air flowrates and biodegradation and volatilization was made. It was found that extracting air at the optimal rate for biodegradation resulted in 90% removal by biodegradation and 10% removal by volatilization. It was also found that passing the contaminants volatilized in the off-gas through clean soil resulted in complete biodegradation of the volatilized vapors.

In situ respiration tests documented that oxygen consumption rates followed zero-order kinetics and that rates were linear down to 2 to 4% oxygen. Therefore, air flowrates can be minimized to maintain oxygen levels between 2 and 4% without inhibiting biodegradation of fuel, with the added benefit that lower air flowrates will increase the percent of removal by biodegradation and decrease the percent of removal by volatilization.

The study was terminated because the process monitoring objectives had been met; biodegradation was still vigorous. Although the TPH had been reduced by only 40% by the time of study termination, the low-molecular-weight aromatics — the BTEX components — were reduced by more than 90% (Figure 3-18). It appeared that the bioventing process more rapidly removed the BTEX compounds than the other JP-4 fuel constituents.

Results from this study demonstrated the effectiveness of bioventing for remediating fuel-contaminated soils, the ineffectiveness of moisture or nutrient addition for increasing in situ biodegradation rates, and the importance of air flowrates for optimizing biodegradation over volatilization. However, it was evident from this study that a long-term bioventing study was necessary to examine process variables. This led to the initiation of the Site 280, Hill AFB and the Site 20, Eielson AFB projects described in the following sections.

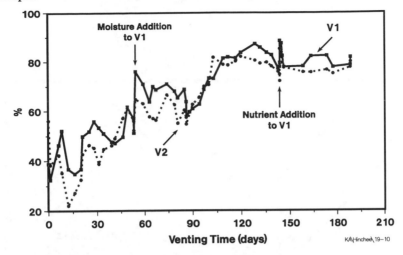

Figure 4-3. Cumulative Percent Hydrocarbon Removal and the Effect of Moisture and Nutrient Addition at Tyndall AFB, Florida

Further details of the Tyndall AFB study may be found in Miller (1990) and Miller et al. (1991).

III. SITE 280, HILL AFB, UTAH

A key objective of the study at Site 280 was to optimize the injection air flowrates. These efforts were intended to maximize biodegradation rates in JP-4 jet fuel-contaminated soils while minimizing or eliminating volatilization. The site studied was a JP-4 jet fuel spill at Hill AFB that had existed since sometime in the 1940s (Figure 4-4). The geology was similar to that of Site 914, but the average contaminant levels were slightly higher (Figure 4-5). Vent wells were installed to a depth of approximately 110 ft and groundwater was at a depth of approximately 100 ft.

From November 1992 through January 1995, a number of studies were conducted to evaluate low-intensity bioremediation at Site 280. These efforts included (1) varying the air injection flowrates in conjunction with in situ respiration tests, and (2) surface emissions testing to provide information for system optimization.

Five air flowrate evaluations were conducted at Site 280 from 1991 through 1994 (28, 67, 67, 40, and 117 cfm [790, 1,900, 1,900, 1,100, and 3,300 L/min]). Each evaluation was followed by in situ respiration testing. The 67 cfm (1,900 L/min) study was repeated to include additional soil gas monitoring points added to the site. Monthly soil gas monitoring was conducted at Site 280 to measure the concentrations of oxygen, carbon dioxide, and TPH at each sampling point following system operation at each of the different air flowrates.

54

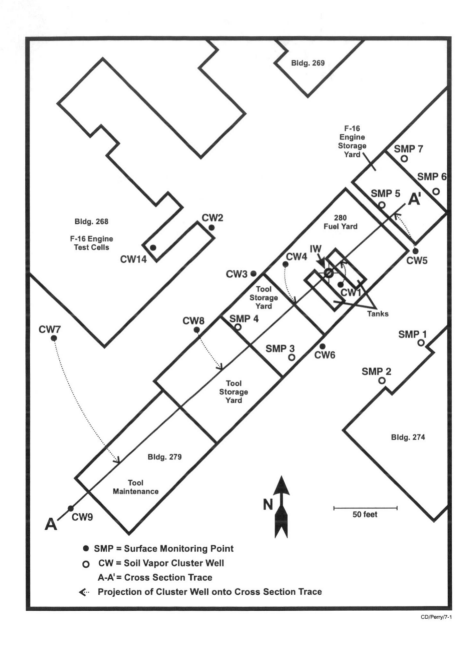

Figure 4-4. Schematic Diagram Showing Locations of Soil Gas Monitoring Points,
Surface Monitoring Points, and Injection Wells at Site 280, Hill AFB, Utah

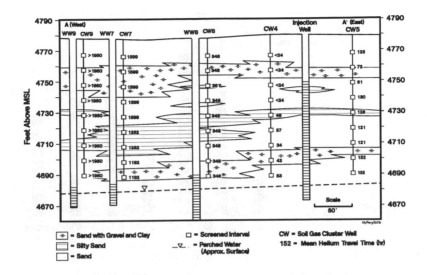

Figure 4-5. Geologic Cross Section Showing Known Geologic Features and Soil TPH
Concentrations (mg/kg) at Site 280, Hill AFB, Utah

Surface emissions tests were conducted during each air injection test and
while the air injection system was turned off. In each of the surface
emissions tests, no significant differences were found between the periods
of air injection and no air injection. TPH soil gas levels measured during
the air injection periods averaged approximately 70 ppmv, while TPH soil
gas levels during resting periods averaged 42 ppmv. These averages were
not found to be statistically different. Likewise, surface emissions rates
were not significantly different at different flowrates.

Final soil sampling was conducted in December 1994. Results from the
initial and final BTEX and TPH samples are shown in Figures 4-6 and 4-7,
respectively. Results shown represent soil samples within a 0- to 25-ft
radius of the injection well and a 25- to 75-ft radius. In general, BTEX
and TPH concentrations decreased at all depths within the 25-ft radius from
the vent well, with the exception of the samples collected at a depth of 90
to 100 ft. Samples taken from this depth are located at the capillary fringe,
and it is likely that adequate aeration was not possible at that location.
Samples collected beyond the 25-ft radius were less conclusive, indicating
that this area was not aerated.

Further details of the Site 280, Hill AFB study may be found in Sayles
et al. (1994b).

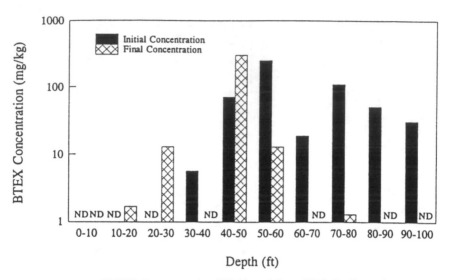

a.   BTEX Concentration Within a 25- to 75-ft Radius of the Injection Well

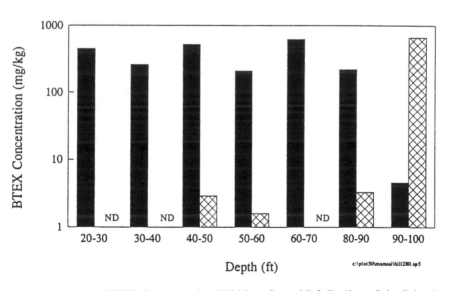

b.   BTEX Concentration Within a 0- to 25-ft Radius of the Injection Well

Figure 4-6.   Site Average Initial and Final BTEX Soil Sample Results at Site 280, Hill
AFB, Utah

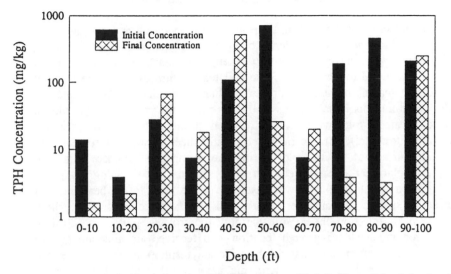

a.   TPH Concentration Within a 25- to 75-ft Radius of the Injection Well

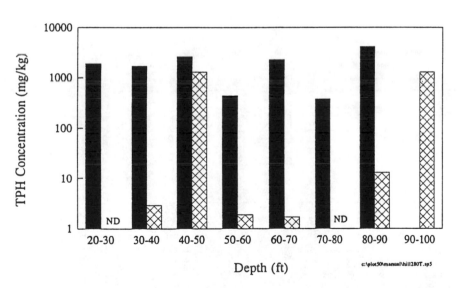

b.   TPH Concentration Within a 0- to 25-ft Radius of the Injection Well

Figure 4-7. Site Average Initial and Final TPH Soil Sample Results at Site 280, Hill
AFB, Utah

58

IV. SITE 20, EIELSON AFB, ALASKA

The objective of the Eielson AFB study was to install and operate an in situ soil bioremediation system to investigate the feasibility of using bioventing to remediate JP-4 jet fuel contamination in a subarctic environment and to actively increase soil temperature to determine to what degree increased soil temperature can enhance the biodegradation rates of JP-4 contaminants in soil. This study comprised four test plots: (1) one in which heated groundwater was circulated through the test plot (active warming test plot); (2) one in which plastic sheeting was placed over the ground surface of the test plot during the spring and summer months to capture solar heat and passively warm the soil (passive warming test plot); (3) one in which heat tape was installed in the test plot to heat the soil directly (surface warming test plot); and (4) a control test plot, which received air injection but no soil warming (control test plot). In addition, an uncontaminated background location also received air injection but no soil warming to monitor natural background respiration rates. The site soils were a sandy silt, with increasing amounts of sand and gravel with depth. Groundwater was typically at approximately the 7-ft depth. Figure 4-8 illustrates site geologic features and typical construction details of site installations.

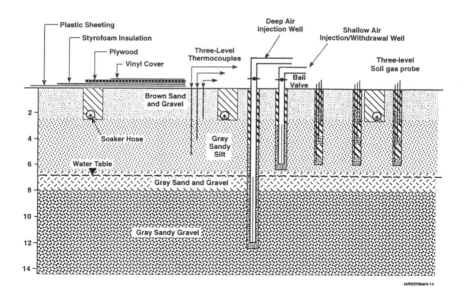

Figure 4-8. Cross Section Showing Geologic Features and Typical Construction Details of the Active Warming Test Plot, Site 20, Eielson AFB, Alaska

Differences in soil temperatures were significant among the four test plots (Figure 4-9). When in operation, the active warming test plot consistently maintained higher temperatures than the other test plots during the winter months. In the passive warming test plot, plastic sheeting increased soil temperature, with average soil temperatures as high at 18°C (64.4°F) during the summer months, compared to average temperatures of approximately 10°C (50°F) in the control test plot. A significant feature of this soil warming technique was that the addition of plastic sheeting in the spring caused a rapid increase in soil temperature, nearly 6 to 8 weeks sooner than in the unheated test plots. The early heating significantly increased the period of rapid microbial degradation. During the winter months, the passive warming test plot remained warmer than the control test plot.

Respiration rates were measured quarterly in each test plot. Of particular interest were rates measured in the control test plot. It was expected that no substantial microbial activity would occur during the winter months in unheated test plots due to the extreme temperatures. However, significant microbial activity was consistently measured in the control test plot, even at soil temperatures just below freezing (Figure 4-10). Respiration rates in the passive warming test plot were observed to increase nearly one order of magnitude as soil temperature increased during the summer months, indicating the success of the use of plastic sheeting to promote soil warming (Figure 4-10). Respiration rates measured in the active warming test plot were higher than those measured in the passive warming or control test plot when warm water circulation was operating. Warm water circulation was discontinued in fall 1993, and as the soil temperature dropped, no significant microbial activity could be measured in the test plot during the winter months. This phenomenon is interesting in that it suggests that during the 2 years of soil heating, microorganisms adapted to growth at higher temperatures, yet lost the ability to remain active in colder soils. In order to determine whether the microbial population could adapt to cold temperatures given time, a final in situ respiration test was conducted in January 1995. Significant microbial activity was measured, comparable to the control test plot, indicating either readaptation or recolonization by the microbial population.

The surface warming test plot has shown promise as a form of soil warming. Soil temperatures and respiration rates were higher than temperatures or rates in either the passive warming or control test plot and were similar to those measured in the active warming test plot during warm water circulation. These results indicate that the use of heat tape may prove to be a more efficient means of soil warming than hot water circulation, because the problem of high soil moisture content is avoided.

An evaluation of cost versus remediation time was conducted to evaluate the feasibility of soil warming. Costs for the basic bioventing system in

60

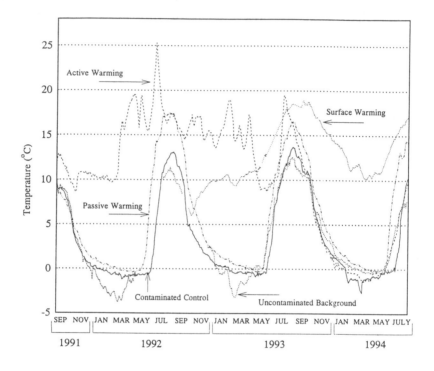

Figure 4-9. Soil Temperature in Four Test Plots and the Background Area at Site 20, Eielson AFB, Alaska

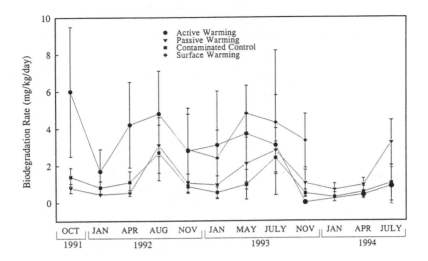

Figure 4-10. Biodegradation Rates in Four Test Plots at Site 20, Eielson AFB, Alaska

Table 4-1 were based on costs calculated by Downey et al. (1994b). Given that average biodegradation rates were higher in the actively warmed plots, overall remediation time would be more rapid than in the unheated test plots (Table 4-1). Although capital costs were higher in the active and surface warming test plots, the rapid remediation time results in lower total costs for power and monitoring. Final costs based on $/yd^3 bioremediated illustrate that costs are comparable between the four treatment cells. These results indicate that implementation of a soil warming technology over basic bioventing is not necessarily based on cost, but on desired remediation time and funds available for operation and maintenance relative to capital costs.

Final soil sampling at this site was conducted in August 1994. Results of initial and final BTEX and TPH samples are shown in Figures 4-11 and 4-12, respectively. A dramatic reduction in BTEX was observed at all sample locations, while TPH on average was reduced by approximately 60%.

Spatial variability in contaminant distribution and biodegradation rates makes quantitative comparison between the test plots difficult; however, the results from the active, surface, and passive warming test plots clearly demonstrate that these forms of soil warming have increased biological activity in these areas. In the active and surface warming test plot, despite problems due to high soil moisture content, biodegradation rates consistently have been higher than those measured in either the passive

Table 4-1
Cost Analysis of Soil Warming Techniques at
Site 20, Eielson AFB, Alaska[1]

Task	Basic Bioventing (no warming)	Active Warming	Passive Warming	Surface Warming
Site Visit/Planning	5,000	5,000	5,000	5,000
Work Plan Preparation	6,000	6,000	6,000	6,000
Pilot Testing	27,000	27,000	27,000	27,000
Regulatory Approval	3,000	6,000	3,000	3,000
Full-Scale Construction				
Design	7,500	7,500	7,500	7,500
Drilling/Sampling	15,000	20,000[2]	15,000	15,000
Installation/Startup	4,000	26,000	10,500	13,000
Remediation Time Required[3]	9.4 years	2.8 years	6.9 years	3.4 years
Monitoring	30,550	9,800	24,150	11,050
Power	13,160	9,800	9,660	17,000
Final Soil Sampling	13,500	13,500	13,500	13,500
Cost per yd^3	$25.50	$26.12	$24.86	$24.21

[1] Costs are estimated based on a 5,000-yd^3 contaminated area with an initial contamination level of 4,000 mg/kg. [2] Requires installation and development of one well. [3] Estimated based on average biodegradation rates in each test plot.

warming or the control test plot, even though the control test plot appears to be more heavily contaminated than the active warming test plot. These results have demonstrated the feasibility of bioventing in a subarctic climate and the potential advantages of soil warming to accelerate remediation.

Further details of the Site 20, Eielson AFB study may be found in Leeson et al. (1995) and Sayles et al. (1994a).

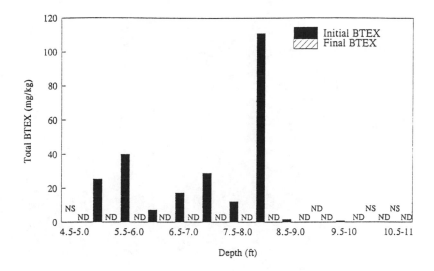

Figure 4-11. Site Average Initial and Final BTEX Soil Sample Results at Site 20, Eielson AFB, Alaska

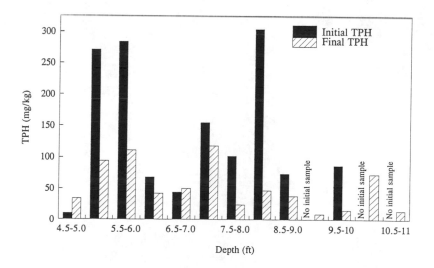

Figure 4-12. Site Average Initial and Final TPH Soil Sample Results at Site 20, Eielson AFB, Alaska

CHAPTER 5

BIOVENTING IMPLEMENTATION:
SITE CHARACTERIZATION ACTIVITIES

Site characterization is an important step in determining the feasibility of bioventing and in providing information for a full-scale bioventing design. This chapter discusses site characterization methods that are recommended for bioventing sites based on field experience and a statistical analysis of the U.S. Air Force Bioventing Initiative data. These parameters have proven to be the most useful in predicting the potential applicability of bioventing at a contaminated site. Figure 5-1 summarizes the sequence of events for characterization of a typical site. Each step presented in Figure 5-1 is discussed in the following sections.

Site characterization activities to be conducted at a potential bioventing site are described in this section as follows:

1. Review existing site data.
2. Conduct soil gas survey.
3. Perform soil gas permeability testing.
4. Perform in situ respiration testing.
5. Characterize soil.

I. EXISTING DATA AND SITE HISTORY REVIEW

The first step in designing and installing a bioventing system is to review the existing site data. This review will provide preliminary information for determining whether bioventing is a feasible option for a specific site. Also, the initial data review will help to identify any additional information that will be needed to complete the bioventing design.

Information to be obtained during the data review, if possible, should include the following:

- types of contaminants;
- quantity and distribution of free product (if present);
- historic water table levels;
- three-dimensional distribution of contaminant;
- potential for a continuing source due to leaking pipes or tanks;
- particle size distribution or soil gas permeability; and
- surface features such as concrete or asphalt.

65

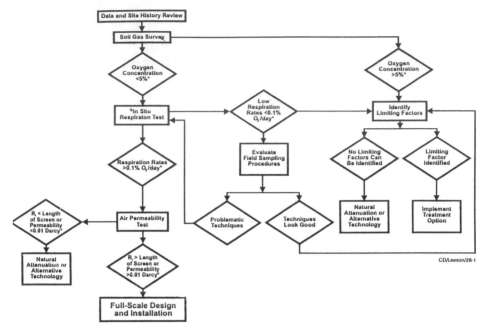

Figure 5-1. Conceptual Decision Tree for Determining the Potential Applicability of Bioventing at a Contaminated Site

At this stage, the most important data is type of contaminant. Bioventing is applicable only to compounds that are biodegraded aerobically, such as petroleum hydrocarbons[1]. Compounds such as chlorinated solvents tend to degrade more readily under anaerobic conditions. At most sites where bioventing is applicable, the contaminant will be petroleum hydrocarbons; however, bioventing also may potentially be applied at some sites contaminated with both chlorinated solvents and petroleum hydrocarbons.

If significant free product is present, removal must be addressed either before or simultaneously with bioventing. Bioventing alone is not sufficient to remediate sites with large quantities of free product. The bioslurping technology combines bioventing and free product removal and is currently under development by the U.S. Air Force (Kittel et al., 1995).

Data on historical water table levels also are important to determine whether contamination is accessible for bioventing or is present below the water table. If significant contamination is present below the water table, dewatering may be needed to complete site remediation. At some sites, bioventing may be feasible only during periods of seasonal low water

---

[1] Refer to Chapter 3 for a discussion of compounds degraded through bioventing.

tables. Developing a three-dimensional distribution of the contaminant will provide information necessary for generating an initial estimate of the screen depths and size of the bioventing system that will be required. This initial estimate will provide guidelines for conducting the soil gas survey and for collecting initial soil samples necessary to estimate the initial mass of contamination at the site.

The potential for a continuing source of contamination must be addressed at any site. Contaminated sites often are created by leaking underground pipes or tanks. These sources must be eliminated if bioventing is to achieve cleanup.

If available, data on particle-size distribution or permeability are useful for determining the potential for applying bioventing. Because the success of bioventing depends on the ability to move air through the soil, particle-size or permeability measurements are critical parameters. However, unless these values are extreme (e.g., saturated clay), initial treatability studies should be conducted to determine bioventing applicability.

Example 5-1 illustrates review and evaluation of existing site data.

Example 5-1. Review of Existing Data and Site History: Application of bioventing is being considered at Area of Concern (AOC) A, Keesler Air Force Base (AFB). The following information is available:

- The soil was contaminated by leakage from underground gasoline storage tanks.
- Storage tanks were removed in 1991.
- A site map (Figure 5-2) was provided with limited TPH soil sample results.
- The soils are very sandy.

After examining the existing site data, we can conclude the following:

- The type of contaminant is gasoline, a very good candidate for bioventing. Based on this information, a soil gas survey is scheduled.
- No information was provided on free product or on water levels. Given that there are groundwater monitoring wells shown in Figure 5-2, it is likely that some information exists. Therefore, we will attempt to find the additional information but will plan also to collect free product and water level measurements during the soil gas survey phase.
- The quantity of the release is unknown because contamination occurred over a long period of time. However, the limited soil sampling provides a general guideline for the area in which to conduct a soil gas survey.
- Because the storage tanks were removed, a continuing source of contamination is not a factor.
- Particle size distribution is known. Soils are sandy, making this site an excellent candidate for bioventing.

68

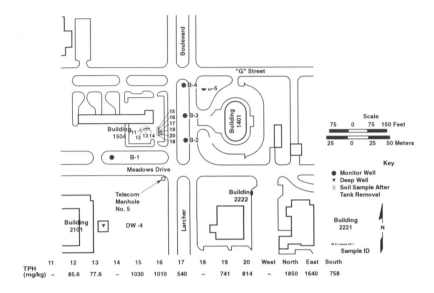

Figure 5-2. Site Map Showing Well Locations and TPH Soil Concentrations at AOC
A, Keesler AFB, Mississippi

II. SOIL GAS SURVEY

At sites where the contamination is at sufficiently shallow depths
(typically <20 ft [6.1 m]), a soil gas survey should be conducted initially
to determine whether oxygen-limited conditions exist. Oxygen-limited
conditions are a good indicator that bacteria capable of degrading the
contaminants of concern are present, given that soil gas in uncontaminated
vadose zone soils generally will exhibit oxygen concentrations equivalent to
ambient air. The soil gas survey also assists in delineating the extent of
contamination and locating suitable areas for vent well and monitoring
point placement. Data on soil gas concentrations of oxygen, carbon
dioxide, and TPH can provide valuable insight into the extent of subsurface
contamination and the potential for in situ bioventing. The procedures
outlined in this section will assist in the collection and interpretation of soil
gas information, with the ultimate goal of promoting a more cost-effective
approach to fuel-contaminated soil remediation.

A. SOIL GAS CHEMISTRY

The chemical composition of soil gas can vary considerably from
ambient air as a result of biological and mineral reactions in the soil.
Many compounds and elements may be present in soil gas due to site
specific geochemistry, but three indicators are of particular interest for
bioventing systems: respiration gases (oxygen and carbon dioxide) and

hydrocarbon vapors. The soil gas concentrations of these indicators in relation to atmospheric air and uncontaminated background soils can provide valuable information on the ongoing natural biodegradation of hydrocarbon contaminants and the potential for bioventing to enhance the rate of natural biodegradation.

1. Respiration Gases

Oxygen serves as a primary electron acceptor for soil microorganisms that degrade both refined and natural hydrocarbons. Following a hydrocarbon spill, if active microbial populations are present, soil gas oxygen concentrations are usually low (typically less than 5%) and soil gas carbon dioxide (a metabolite of hydrocarbon degradation) may be high (typically >10%). Oxygen concentrations generally are lower in the vicinity of the contaminated soils than in clean soils, indicating that aerobic biodegradation is depleting oxygen. As the population of fuel-degrading microorganisms increases, the supply of soil gas oxygen often is depleted, creating an anaerobic volume of contaminated soil. Under anaerobic conditions, fuel biodegradation generally proceeds at significantly slower rates than when oxygen is available for metabolism. In some cases, aerobic biodegradation will continue because the diffusion or advection of oxygen into soils from the atmosphere exceeds biological oxygen utilization rates. Under these circumstances, the site is naturally aerated, and the hydrocarbons will be naturally attenuated over time.

Carbon dioxide is produced as a by-product of the complete aerobic biodegradation of hydrocarbons and also can be produced or buffered by the soil carbonate cycle (Ong et al., 1991). Carbon dioxide levels in soil gas generally are elevated in fuel contaminated soils when compared to levels in clean background soils. In many soils, higher carbon dioxide concentrations correlate with low oxygen levels; however, this is not always true. Due to the buffering capacity of alkaline soils, the relationship between contaminant biodegradation and carbon dioxide production is not always a reliable indicator. Carbon dioxide can form carbonates rather than gaseous carbon dioxide, particularly in soils with pH over 7.5 and high reserve alkalinity. In acidic soils, such as exist at Tyndall AFB, Florida, carbon dioxide production is directly proportional to oxygen utilization (Miller and Hinchee, 1990).

It is important to compare soil gas survey results for a contaminated area with those obtained from an uncontaminated area. Typically, soil gas concentrations in an uncontaminated area will be significantly different, with oxygen concentrations approximately equal to ambient concentrations and very low carbon dioxide concentrations (<0.5%).

2. Hydrocarbon Vapors

Volatile hydrocarbons found in soil gas can provide valuable information on the extent and magnitude of subsurface contamination. Fuels such as gasoline, that contain a significant fraction of C_6 and lighter compounds, are easily detected using soil gas monitoring techniques. Heavier fuels, such as diesel, contain fewer volatiles and are more difficult to locate through volatile hydrocarbon monitoring. Methane frequently is produced as a by-product of anaerobic biodegradation and, like oxygen depletion, has been used to locate the most contaminated soils at a site. Extensive literature is available on soil gas survey techniques that use volatile hydrocarbons as indicators of contamination (Rivett and Cherry, 1991; Downey and Hall, 1994).

B. COLLECTION AND ANALYSIS OF SOIL GAS SAMPLES

This section describes the test equipment and methods used to conduct field soil gas surveys, to monitor soil gas for bioventing systems, and to install temporary soil gas monitoring points. The procedures and equipment described in this section are intended as guidelines. Because of widely varying site conditions, site-specific applications will be required. In some regulatory jurisdictions, soil gas survey monitoring points must comply with well-installation or other regulations.

Whenever possible, soil gas surveys should be conducted at potential bioventing sites before the pilot test vent well(s) and monitoring points are situated. The soil gas survey is used to determine if bioventing is required based on whether anaerobic soil gas conditions exist, and to provide an initial indication of the extent of contamination. If sufficient oxygen is naturally available and distributed throughout the subsurface, bioventing may not enhance biodegradation rates. The soil gas survey can help to determine the areal extent and, in the case of shallow contamination, the vertical extent of soil contamination. Information about contaminant distribution helps to locate the vent well and soil gas monitoring points and to determine the optimum depths of screened intervals.

The soil gas survey points should be arranged in a grid pattern centered on the known or suspected contaminated area. The soil gas probes are positioned at each grid intersection, and the survey begins near the center of the grid and progresses outward to the limits of significant detectable soil contamination. At some sites, soil gas measurements can be taken at a number of depths at each location to determine the vertical distribution of contamination and oxygen supply. At shallow sites, a soil gas sampling grid should be completed with samples collected from multiple depths if the contaminated interval exceeds 3 ft (0.91 m) or if contamination is suspected in different soil types.

A soil gas survey can be conducted using small-diameter (typically ⅝- to 1-inch [1.6 to 2.5 cm] outside diameter [OD]) steel probes. The typical probe consists of a drive point with a perforated tip that is threaded onto a series of drive rod extensions. Figure 5-3 shows a typical setup for monitoring soil gas[1].

The method of probe installation will be dictated by soil conditions and depth of contamination. Utility clearances from the local utility companies and digging permits (required at military installations) should be obtained prior to probe installation. Temporary probes are installed using either a hand-held electric hammer or a hydraulic ram. The maximum depth for hammer-driven probes is typically 10 to 15 ft (3 to 4.6 m), depending on soil texture. Hydraulic rams are capable of driving the probes over 30 ft (9.1 m) in a variety of soil conditions. If hydraulic rams are not sufficient, a Geoprobe™ or similar equipment can be used to drive the probe and also to collect soil samples.

At sites with deeper contamination where soil texture precludes the use of hammer or hydraulic ram or where a permanent monitoring system is required, permanent soil gas monitoring points can be installed using either a portable or a truck-mounted drill rig.

Gaseous concentrations of carbon dioxide and oxygen can be analyzed using an O_2/CO_2 analyzer. The analyzer generally will have an internal battery-powered sampling pump and range settings of 0 to 25% for both oxygen and carbon dioxide. Prior to taking measurements, the analyzer should be checked for battery charge level and should be calibrated daily using atmospheric concentrations of oxygen and carbon dioxide (20.9 and 0.05%, respectively) and a gas standard containing 0.0% oxygen, 5.0% carbon dioxide, and 95% nitrogen.

Several types of instruments are available for field measurement of TPH concentrations in air. The selected instrument must be able to measure hydrocarbon concentrations in the range of 1 to 10,000 parts per million, volume per volume (ppmv) and be able to distinguish between methane and nonmethane hydrocarbons. Flame ionization detectors are the most accurate field screening instruments for fuel hydrocarbons. Instruments that use a platinum catalyst detector system also are acceptable and are easier to use in the field. Photoionization detectors are not recommended for the high levels of volatile hydrocarbons found at many sites. Before measurements are taken with any field instrument, the battery charge level should be checked and the analyzer should be calibrated against a hexane calibration gas to ensure proper operation.

The analyzer should also have a selector switch to change the response to eliminate the contribution of methane gas to the TPH readings. Methane gas is a common constituent of anaerobic soil gas and is generated by

---

[1] Refer to Appendix B for recommended specifications and manufacturers for soil gas sampling equipment.

72

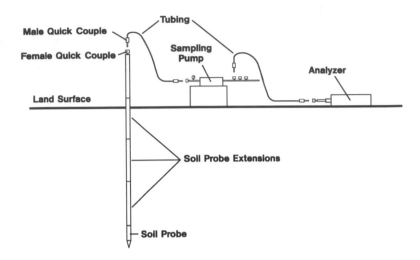

Figure 5-3. Schematic Diagram of a Soil Gas Sampling System Using the Stainless Steel Soil Gas Probe

degrading manmade hydrocarbons or natural organics. Methane is commonly produced in swampy areas or in fill areas containing organic material. If the methane is not excluded from the TPH measurement, TPH results may indicate erroneously high levels of petroleum hydrocarbon contamination in the soil. The methane content can also be estimated by placing a large carbon trap in front of the hydrocarbon analyzer. Heavier hydrocarbons will be retained by the carbon trap while methane and other lighter-molecular-weight hydrocarbons pass through to the detector.

Electric motor-driven sampling pumps are used to both purge and collect samples from monitoring points and soil gas probes. The pumps should be either oil-less rotary-vane or diaphragm pumps capable of delivering approximately 1 cfm (28 L/min) of air at a maximum vacuum of 270"H$_2$O (6.7 × 10^4 Pa). The pumps have oil-less filters to eliminate particulates from the air stream. Low-flow battery-operated pumps may be favored in high permeability soils to minimize short-circuiting.

Differential vacuum gauges are used to monitor the vacuum in the sampling point during purging and as an indicator of relative permeability. Typical vacuum ranges of the gauges are 0 to 50"H$_2$O (0 to 1.2 × 10^4 Pa) and 0 to 250"H$_2$O (6.2 × 10^4 Pa) for sites with sandy and clayey soils, respectively.

Purging the soil gas probe is a prerequisite for obtaining representative soil gas samples. A typical purging system consists of a 1-cfm (28 L/min) sampling pump, a vacuum gauge, and an O$_2$/CO$_2$ meter. The vacuum side of the pump is connected to the soil gas probe. A vacuum gauge is attached to a tee in the vacuum side of the system to monitor the vacuum produced during purging, and the O_2/CO_2 analyzer is connected to a tee in the outlet tubing to monitor O_2/CO_2 concentrations in the extracted soil gas. The magnitude of vacuum measured during purging is inversely proportional to soil permeability and will determine the method of sample collection.

After the purging system is attached to the soil gas probe or monitoring point, the valve or hose clamp is opened and the pump is turned on. Purging is continued until oxygen and carbon dioxide concentrations stabilize, indicating the purging is complete. Before the pump is turned off, a hose clamp or valve is used to close the sampling tubing to prevent fresh air from being drawn into the soil gas probe.

Sampling methods for high-permeability soils (sand and silt) should be followed if the vacuum measured during purging is less than $10''H_2O$ (2.5 $\times$ 10^3 Pa). Soil gas sampling and analysis are performed using the same equipment used for purging, minus the vacuum gauge. After opening the sampling point valve or hose clamp is opened, the sampling pump is turned on and the extracted soil gas is analyzed for stable oxygen/carbon dioxide and TPH concentrations.

A different sampling procedure can be followed to collect soil gas samples from low-permeability soils. The higher vacuums required for sampling increase the risk of vacuum leaks that would introduce fresh air and dilute the soil gas sample. One method which may be utilized in low-permeability soils is described in the following paragraph.

After the sampling point is purged, a soil gas sample is collected in a Tedlar® bag which is inside an air-tight chamber. The chamber is connected to the sampling point via a hose barb that passes through the chamber wall and then closed, sealed, and connected to the pump inlet with flexible tubing. The sampling system is shown in Figure 5-4. To collect the sample, the monitoring point valve is opened, the pump is turned on, and the pressure relief port on the chamber is sealed with either a valve or the sampler's finger. The partial vacuum created by the pump within the chamber will draw soil gas into the Tedlar® bag. When the Tedlar® bag is nearly filled, the sampling point valve or hose clamp is closed, and the pump is turned off. Then the chamber is opened, the Tedlar® bag valve is closed, and the bag is removed from the chamber. The soil gas sample is analyzed by attaching the O_2/CO_2 and TPH analyzers directly to the Tedlar® bag. The advantage of this method is that the sampling pump is no longer in line, thereby shortening the sampling train and minimizing subsequent sample dilution.

Most problems encountered during soil gas sampling and purging can be divided into three categories: (1) difficulty extracting soil gas from the sampling point, (2) water being drawn from the sampling point, and (3)

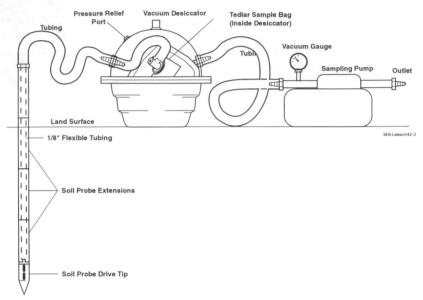

Figure 5-4. Schematic Diagram of a Soil Gas Sampling System for Collection of Soil Gas from Low Permeability Soils

high oxygen readings in areas of known soil contamination. Some of the more common problems and solutions are discussed below.

Difficulty extracting soil gas from a sampling point typically is caused by low-permeability (clayey and/or nearly saturated) soils. Collecting soil gas samples from low-permeability soils is facilitated by slowing the soil gas extraction rate, thus allowing the use of less vacuum. Difficulty extracting soil gas from a soil gas probe can be caused also by the screen being fouled by fine-grained soil or heavy petroleum residuals. The probe should be removed from the soil, and the screen should be either cleaned or replaced if visibly fouled.

Water being drawn from the sampling point by the purge pump can be the result of either the point being installed in the saturated zone or, in the case of permanent monitoring points, the filter pack being saturated with water during construction. In the former case, a temporary probe can be pulled up to a shallower depth above the saturated zone and resampled. With a permanent monitoring point installed within the saturated zone, sampling must be delayed until either the water table drops because of seasonal variations or the water table is artificially depressed by a dewatering operation.

If the screened interval in a permanent monitoring point is installed above the saturated zone but the filter pack was saturated with water during construction, sampling may still be possible if the water is pumped from the monitoring point. This method will work only if the screened interval is at a depth of less than approximately 22 ft (6.7 m), which is the practical limit of suction lift.

Water also can be drawn into the point in unsaturated soils if a vacuum in excess of capillary pressure is created. In this case, the extracted flow typically is a mixture of water and soil gas. Frequently, a water trap can be used before the sampling pump to remove the water and make it possible to collect and analyze a soil gas sample.

High soil gas oxygen readings in areas of known soil contamination may indicate a leak in the sampling or purging system. The potential for leakage, and the resulting dilution of the sample with atmospheric air, is higher in low-permeability soils where higher vacuums are required for purging and sampling. If a leak is suspected, all connections in the sampling system and the seal around the monitoring point or soil gas probe should be inspected for leaks. Seals around a soil gas probe or monitoring point can be checked for leaks by inspecting for air bubbles while injecting air with a sampling pump after adding water around the probe or monitoring point. Any observed or suspected leaks should be corrected by tightening connections, repositioning the soil gas probe, or attempting to repair the monitoring point seal.

C. INTERPRETATION OF SOIL GAS SURVEY RESULTS

The purpose of gathering soil gas data during bioventing investigations is to locate areas where addition of oxygen will most efficiently enhance fuel biodegradation. Low soil gas oxygen concentrations are a preliminary indication that bioventing may be feasible at the site and that it is appropriate to proceed to in situ respiration testing. If soil gas oxygen concentrations are high (>5 to 10%), but contamination is present, other factors may be limiting biodegradation. The most common limiting factor is low moisture level. If a pilot test is to be completed, the soil gas survey should focus on locating areas having the lowest oxygen concentrations. For full-scale applications, it is useful to determine the entire areal extent and depth of soils with an oxygen deficit (for practical purposes less than 5% oxygen).

Diffusion, biometric pumping, or water table fluctuations can enhance air movement into very shallow, permeable soils and provide a natural oxygen supply. Soil gas data are useful for determining which sites are naturally aerated and therefore do not require mechanical bioventing systems.

If high oxygen concentrations are observed on the site, the existence of significant contamination is questionable. It is possible that lower levels of contamination (i.e., <1,000 mg/kg TPH) could be biodegraded by the natural oxygen supply and no active remediation would be necessary. If higher levels of hydrocarbons are present (≥1,000 mg/kg), it is unlikely that the natural oxygen supply is adequate to sustain biodegradation;

76

therefore, it is likely that some other factor is limiting biodegradation. In the authors' experience, soil containing both high oxygen and high hydrocarbon concentrations only occur at moisture-limited sites (the most common case) or sites with toxicity problems (TCE in one case and phenolics in another). The authors are aware of only two cases where the lack of oxygen utilization was not explained by these factors. These occurred at a JP-5 jet fuel site on Fallon Naval Air Station (NAS) in Nevada and at a JP-4 spill site at Davis-Monthan AFB in Arizona. The problem sites are not moisture-limited, and to date no clear explanation has arisen (Engineering-Science, 1994a; Kittel et al., 1994). A series of examples of soil gas survey results and data interpretations is presented here to illustrate the principles discussed in this section.

Example 5-2. Soil Gas Survey Conducted at Keesler AFB: At the site described in Example 5-1, a soil gas survey was to be conducted. First, depth to groundwater and free product thickness were measured at all of the groundwater monitoring wells. Groundwater depths were as follows: MW8-1 at 6.8 ft (2.1 m), MW8-2 at 8.0 ft (2.4 m), MW8-3 at 8.2 ft (2.5 m), and MW8-11 at 8.25 ft (2.5 m). No free product was detected in any of the wells, so free product removal is not a factor at this site.

A limited soil gas survey was conducted at this site since the area of contamination had recently been defined. Soil gas samples were collected at depths ranging from 2 to 6 ft (1.6 to 1.8 m). Because groundwater was measured at 6.8 ft (2.1 m), soil gas probes were not driven deeper.

Results from this survey are shown in Table 5-1. At most locations, oxygen was limiting with concentrations less than 5% and carbon dioxide and TPH concentrations were relatively high. The exception was at location SGS-D-6.0'. At this point, oxygen was measured at 20.1%, carbon dioxide at 0.1%, and TPH at 120 ppm.

These levels were more representative of ambient air than of the soil gas concentrations measured at other points at the site, indicating there may be significant dilution of this sample. Because of these measurements, the sampling pump was thoroughly examined and loose connections were tightened. Upon resampling, soil gas concentrations were more representative of other soil gas concentrations. If resampling had produced the same initial results, it could be possible that this monitoring point was plugged, causing the sampling train to leak, and/or that atmospheric air was short circuiting to the point. In either case, results from this point should be discarded as invalid.

Results of this soil gas survey indicate that this site is an excellent candidate for bioventing.

Table 5-1
Results from a Soil Gas Survey at AOC A, Keesler AFB, Mississippi

Soil Gas Survey (SGS) Point	Depth (ft)	Oxygen (%)	Carbon Dioxide (%)	TPH (ppmv)
SGS-A	2.0	4.8	9.8	>100,000
	4.0	0.3	12	>100,000
	6.0	0.5	11	>100,000
SGS-B	2.0	1.5	12	>100,000
	4.0	0.5	12	>100,000
	6.0	0.9	12	>100,000
SGS-C	2.0	0.4	11	28,000
	4.0	0.8	11	30,000
	6.0	0.4	11	32,000
SGS-D	2.0	0.4	11	47,000
	4.0	0.3	11	56,000
	6.0	20.1	0.1	120
	6.0	0.4	11	60,000

Example 5-3. Soil Gas Survey at the Aquasystem Site, Westover AFB, Massachusetts: This site consisted of USTs which, when removed, revealed soil contamination. An unknown quantity of mixed-fuels contamination remained in the soil. Site soils were predominantly sand, with groundwater at approximately 13 ft (4.0 m) below the surface. A soil gas survey consisting of a 12-point grid was completed in and downgradient of the former tank pit. All points were sampled at multiple depths. Results of the survey are provided in Table 5-2.

Low levels of TPH were detected in the soil gas at this site. Oxygen levels were significantly depleted below atmospheric concentrations in soils near PT7 and PT17 and generally decreased with depth. However, the 8 to 9% of oxygen available in this area is more than sufficient to sustain in situ biodegradation. Carbon dioxide ranged from 2 to 8.5% and generally increased with depth. The available data suggest that significant natural biodegradation is occurring at the site. It is possible that more oxygen-depleted soil exists in the capillary fringe, and that engineered bioventing could accelerate biodegradation if this anaerobic zone exists. The decision to biovent this site should be based on other factors, such as the potential risk that soil contamination poses to groundwater.

Table 5-2
Results from a Soil Gas Survey at the Aquasystem Site, Westover AFB, Massachusetts

Soil Gas Survey Point	Depth (ft)	Oxygen (%)	Carbon Dioxide (%)	TPH (ppmv)
PT1	3.0	16	3.2	60
	6.0	12.5	5	60
PT2	3.0	15.5	4.3	72
	6.0	13	6	74
PT3	3.0	18	2.6	74
	6.0	12	6.2	84
PT4	3.0	16	4	86
	6.0	11.5	5	80
PT5	3.0	14.8	4	76
	6.0	11	5.2	72
PT7	3.0	14	7	105
	6.0	8.5	8.5	69
PT8	3.0	12	5.5	75
	6.0	11	6.5	76
PT9	3.0	11.5	6	90
	6.0	11	6.2	78
PT11	3.0	16	3.5	84
	6.0	15	4	94
PT12	3.0	18.5	2.5	80
	6.0	15.5	4.2	91
	9.0	15	4.8	90
	12.0	13	5.6	92
PT16	6.0	17	2	94
	7.5	13	3.5	80
PT17	6.0	11.8	6.5	92
	9.0	11	6.5	96

III. IN SITU RESPIRATION TESTING

The in situ respiration test was developed to provide rapid field measurement of in situ biodegradation rates. This information is needed to determine the potential applicability of bioventing at a contaminated site and to provide information for a full-scale bioventing system design. This section describes the test as developed by Hinchee and Ong (1992). This in situ respiration test has been used at each of the U.S. Air Force Bioventing Initiative sites and at numerous other sites throughout the United States. The in situ respiration test described in this document is essentially the same as that described by Hinchee and Ong (1992), with minor modifications.

A. IN SITU RESPIRATION TEST PROCEDURES
The in situ respiration test is conducted by placing narrowly screened soil gas monitoring points into the unsaturated zone of contaminated soils and venting these soils for a given period of time with air containing an inert tracer gas (typically helium). The apparatus for the respiration test is illustrated in Figure 5-5[1]. An example procedure for conducting an in situ respiration test is provided in Appendix C.

As part of the U.S. Air Force Bioventing Initiative, respiration rates in uncontaminated areas of similar geology to the contaminated test site were evaluated. Given the results, it was evident that measurement of background respiration rates was not necessary, since there was little significant respiration. Instead, it is recommended that oxygen and carbon dioxide be measured in an uncontaminated location of similar geology, and, if there is significant oxygen depletion, only then should a background in situ respiration test be conducted, since there may be significant background respiration.

In a typical experiment, a cluster of three to four soil gas probes are placed in the contaminated soil of the test location. These soil gas probes must be located in the center of contaminated areas where low soil gas oxygen concentrations and high TPH concentrations have been measured. If the monitoring points are not located in contaminated areas, the in situ respiration test will not produce meaningful results. Additional detail on monitoring point location and construction is provided in Chapter 6.

Measurements of carbon dioxide and oxygen concentrations in the soil gas are taken prior to air and inert gas injection. A 1 to 3% concentration of inert gas is added to the injection air, which is injected for approximately 24 hours at flowrates ranging from 1.0 to 1.7 cfm (28 to 48

---

[1] Refer to Appendix B for recommended specifications and manufacturers for the in situ respiration testing equipment.

80

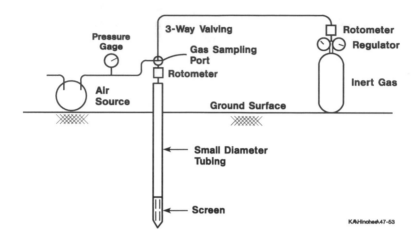

Figure 5-5. In Situ Respiration Test Apparatus

L/min). The air provides oxygen to the soil, and the inert gas measurements provide data on the diffusion of oxygen from the ground surface and the surrounding soil and to ensure that the soil gas sampling system does not leak. The background control location is placed in similar soils in an uncontaminated area to monitor natural background respiration rates.

After air and inert gas injection are turned off, oxygen, carbon dioxide, and inert gas concentrations are monitored over time. Before a reading is taken, the probe is purged for a few minutes until the carbon dioxide and oxygen readings are constant. Initial readings are taken every 2 hours and then progressively over 4- to 8-hour intervals. If oxygen uptake is rapid, more frequent monitoring may be required. If it is slower, less frequent readings may be acceptable. The experiment usually is terminated when the soil gas oxygen concentration is approximately 5%.

As discussed in Section II, at shallow monitoring points there is a risk of pulling in atmospheric air in the process of purging and sampling. Excessive purging and sampling may result in erroneous readings. There is no benefit in oversampling and, when sampling shallow points, care must be taken to minimize the volume of air extraction. In these cases, a low-flow extraction pump of about 0.03 to 0.07 cfm (0.85 to 2.0 L/min) should be used.

B. INTERPRETATION OF IN SITU RESPIRATION TEST RESULTS

Oxygen utilization rates are determined from data obtained during the in situ respiration test. The rates are calculated as the zero order relationship between percent oxygen and time. Typically, a rapid linear decrease in oxygen is observed, followed by a lag period once oxygen concentrations drop below approximately 5%. To calculate oxygen utilization rates, only the first linear portion of the data is used, because this represents oxygen utilization when oxygen is not limiting, as is the case during active bioventing.

To estimate hydrocarbon biodegradation rates from the oxygen utilization rates, a stoichiometric relationship for the oxidation of the contaminant is used. For hydrocarbons, hexane is used as the representative hydrocarbon. If a site is contaminated with compounds other then petroleum hydrocarbons, a suitable compound should be used to determine stoichiometry. The stoichiometric relationship used to determine petroleum degradation rates is:

$$C_6H_{14} + 9.5O_2 \rightarrow 6CO_2 + 7H_2O \qquad (5\text{-}1)$$

Based on the utilization rates (% oxygen per day), the biodegradation rate in terms of mg hexane-equivalent per kg of soil per day is estimated using Equation 5-2:

$$k_B = \frac{\dfrac{k_o}{100}\,\theta_a\,\dfrac{1L}{1{,}000\ cm^3}\,\rho_{O_2}\,C}{\rho_k\left(\dfrac{1\ kg}{1{,}000\ g}\right)} = \frac{-k_o\,\theta_a\,\rho_{O_2}\,C\,(0.01)}{\rho_k} \qquad (5\text{-}2)$$

where: k_B = biodegradation rate (mg/kg-day)

   k_o = oxygen utilization rate (%/day)

   θ_a = gas-filled pore space (volumetric content at the vapor phase, m^3 gas/cm^3 soil)

   ρ_{O2} = density of oxygen (mg/L)

   C = mass ratio of hydrocarbons to oxygen required for mineralization (1/3.5)

   ρ_k = soil bulk density (g/cm^3)

These terms may be derived through either direct measurement or estimation. The oxygen utilization rate, k_o, is directly measured in the in situ respiration test. The ratio of hydrocarbons to oxygen required for mineralization, C, can be calculated based on stoichiometry (see Equation 5-1 for hexane) but generally will fall between 0.29 and 0.33. This neglects any conversion to biomass, which probably is small and difficult, if not impossible, to measure. The density of oxygen may be obtained

82

from a handbook for a given temperature and pressure or calculated from the ideal gas law. Table 5-3 provides some useful oxygen density information. The bulk density of soil is difficult to accurately measure due to the difficulty of collecting an undisturbed sample; however, it may be reasonably estimated from the literature. Table 5-4 lists useful literature values for bulk density.

The gas-filled porosity, θ_a, is the single parameter in Equation 5-2 with the most variability. Theoretically, it can be related to the total porosity, soil bulk density, and moisture content. A doubling of the air-filled porosity results in a doubling of the estimated hydrocarbon degradation rate. Gas-filled porosity may be as high as 0.5 to 0.6 in some very dry clays, but saturated soil is zero. To collect soil gas samples, the gas-filled porosity must be sufficient to allow gas flow. Therefore, it is not possible to conduct an in situ respiration test at very low gas-filled porosity. At most bioventing sites, θ_a ranges from 0.1 to 0.4. Soil in a core or split-spoon sample will be compressed, thereby reducing θ_a. It can be estimated as follows:

$$\theta_a = \theta - \theta_w \tag{5-3}$$

where: θ = total porosity (cm³/cm³)
θ_w = water-filled porosity (cm³/cm³)

The total void volume may be estimated as:

$$\theta = 1 - \frac{\rho_k}{\rho_T} \tag{5-4}$$

where: ρ_k = soil bulk density (g dry soil/cm³) (from Table 5-4)
ρ_T = soil mineral density (g/cm³), estimated at 2.65

The water-filled void volume then can be calculated as:

$$\theta_w = M \frac{\rho_k}{\rho_T} \tag{5-5}$$

where: M = soil moisture (g moisture/g soil)

Table 5-3
Oxygen Density Versus Temperature

Temperature (°C)	Temperature (°F)	Density (mg/L)[1]	Density (lb/ft³)[1]
–3	26.6	1,446[3]	0.090[3]
0	32	1,429[3]	0.089[3]
5	41	1,403[3]	0.088[3]
10	50	1,378[3]	0.086[3]
15	59	1,354[3]	0.084[3]
20	68	1,331[3]	0.083[3]
27	80.6	1,301[2]	0.082[2]
30	86	1,287[3]	0.080[3]
35	95	1,266[3]	0.079[3]
40	104	1,246[2]	0.078[3]

[1] Oxygen density at standard pressure. [2] Density values from Braker and Mossmon, 1980. [3] Density calculated using the second virial coefficient to the equation of state for oxygen gas:

$$P = \frac{RT}{V}\left[1 + \frac{B(T)}{V}\right]$$

where P = pressure (atm), R = gas constant, V = molar volume, and B = second virial coefficient. The temperature dependence of B was calculated from:

$$B(T) = \sum_{i=1}^{n} A_i \left[\frac{T_0}{T} - 1\right]^{i-1}$$

The constants A_i were obtained from Lide and Kehianian (1994).

Table 5-4

Bulk Density of Various Soils[1]

Soil Description	Porosity	Soil Bulk Density, ρ_k (dry g/cm^3)
Uniform sand, loose	0.46	1.43
Uniform sand, dense	0.34	1.75
Mixed-grain sand, loose	0.40	1.59
Mixed-grain sand, dense	0.30	1.86
Windblown silt (loess)	0.50	1.36
Glacial till, very mixed-grained	0.20	2.12
Soft glacial clay	0.55	1.22
Stiff glacial clay	0.37	1.70
Soft slightly organic clay	0.66	0.93
Soft very organic clay	0.75	0.68
Soft montmorillonitic clay (calcium bentonite)	0.84	0.43

[1] From Peck et al. (1962).

Because the water-filled porosity (θ_w) is a difficult parameter to estimate accurately, it frequently is assumed to be 0.2 or 0.3.

Using several assumptions, values for θ_a, ρ_{O2}, C, and ρ_k can be calculated and substituted into Equation 5-2. Assumptions used for these calculations are:

- Gas filled porosity (θ_a) of 0.25
- Soil bulk density (ρ_k) of 1.4 g/cm
- Oxygen density (ρ_{O2}) of 1,330 mg/L
- C, hydrocarbon-to-oxygen ratio of 0.29 from Equation 5-1 for hexane.

The resulting equation is:

$$k_B = \frac{-(k_o)\ (0.25)\ (1,330)\ (0.29)\ (0.01)}{1.4} = -0.68\ k_o \qquad (5-6)$$

The biodegradation rates measured by the in situ respiration test appear to be representative of those for a full-scale bioventing system. Miller (1990) conducted a 9-month bioventing pilot project at Tyndall AFB at the same time Hinchee et al. (1991b) were conducting an in situ respiration test.   The oxygen utilization rates (Miller, 1990) measured from nearby active treatment areas were virtually identical to those measured in the in situ respiration test.   Oxygen utilization rates greater than 1.0%/day are a good indicator that bioventing may be feasible at the site and that it is appropriate to proceed to soil gas permeability testing.   If oxygen utilization rates are less than 1.0%/day, yet significant contamination is present, other factors may be involved in limiting biodegradation.   In this case, other process variables as discussed in Section V below should be considered as limiting biodegradation.   Identifying these other process variables may require additional soil sampling and analysis.   If none of these other process variables can be identified as potentially limiting microbial degradation, alternative technologies may have to be employed for site remediation.

Example 5-4. Results From An In Situ Respiration Test Conducted at Keesler AFB:
At the site described in Example 5-1, an in situ respiration test was conducted. After the soil gas survey, three-level monitoring points were installed at each of the soil gas survey point locations, because these areas were highly contaminated and were oxygen limited. Initial soil gas readings were taken at each of the monitoring points and are shown in Table 5-5. Since all locations were oxygen limited, it was decided to inject air at the deepest level of each of the monitoring points (K1-MPA-7.0', K1-MPB-7.0', K1-MPC-7.0', and K1-MPD-7'1").

Table 5-6 contains data collected at each monitoring point during the in situ respiration test. The oxygen utilization rate is determined as the slope of the % oxygen versus time curve. Only data beginning with that taken at t=0 that appear linear with time were used to calculated the slope. A zero-order respiration rate as seen in these data is typical of most sites (Figure 5-6).   Calculated oxygen utilization rates and corresponding biodegradation rates for these data are shown in Table 5-7.

Results of this test indicate that this site is an excellent candidate for bioventing.

Example 5-4 illustrates the calculation of oxygen utilization data that is linear with time. However, in some instances, this relationship will not be linear and only selected data should be used to calculate the oxygen utilization rate. Example 5-5 illustrates calculation of the oxygen utilization rate from nonlinear data.

Example 5-5. Calculation of Oxygen Utilization Rates From Nonlinear Data: Table 5-8 contains sample data from the Solid Waste Management Unit (SWMU) 66, Keesler AFB. The oxygen utilization rate is determined as the slope of % oxygen versus time curve. Only data beginning with that taken at t=0 that appear linear with time should be used to calculate the slope. A fairly rapid change in oxygen levels was observed at Keesler AFB (Figure 5-7). In this case, the oxygen utilization rate was obtained from the initial linear portion of the respiration curve, which included data from t=0 to t=30.5 hr. As shown, after this point, oxygen concentrations dropped below 5%, and were limiting. The calculated oxygen utilization rate was 11%/day.

Table 5-5
Initial Soil Gas Readings at Monitoring Points at AOC A, Keesler AFB, Mississippi

Monitoring Point	Depth (ft)	Oxygen (%)	Carbon Dioxide (%)	TPH (ppmv)
K1-MPA	3.0	0.1	16	> 100,000
	5.0	0.4	15	> 100,000
	7.0	0.6	15	> 100,000
K1-MPB	2.5	0.5	15	> 100,000
	4.0	0.5	15	> 100,000
	7.0	0.8	15	> 100,000
K1-MPC	3.0	0.4	14	28,000
	5.0	0.1	15	30,000
	7.0	0.5	15	29,000
K1-MPD	3.0	0.6	14	45,000
	5.0	0.5	15	54,000
	7'1"	0.5	15	58,000
Background		16.8	4.6	140

Table 5-6
Raw Data From an In Situ Respiration Test at AOC A, Keesler AFB, Mississippi

Time (hr)	K1-MPA-5.0'			K1-MPA-7.0'		
	O_2 (%)	CO_2 (%)	He (%)	O_2(%)	CO_2(%)	He(%)
0	20.7	0	1.4	20.5	0	1.4
5	20.6	0	1.6	20.6	0	1.4
10	20.1	0.1	1.4	20.3	0.1	1.4
25	19.0	0	1.75	20.1	0	1.6
37	17.8	0	1.4	19.5	0	1.4
50	16.9	0.6	1.4	18.7	0.2	1.25
75	15.2	1.2	1.6	17.3	1.2	1.6
99	14.0	2.0	1.4	16.3	1.2	1.4
Time (hr)	K1-MPB-5.0'			K1-MPC-7.0'		
0	20.6	0	1.6	20.8	0	1.3
5	20.2	0	1.8	20.5	0.2	1.5
10	19.4	0	14	20.2	0.2	1.4
25	16.9	0	1.6	19.5	0	1.3
37	14.8	0	1.4	18.1	0.6	1.2
50	12.9	1.0	1.4	16.9	1.5	1.2
75	9.9	2.6	1.2	13.9	3.0	1.0
99	8.0	3.0	1.2	11.0	4.0	1.0

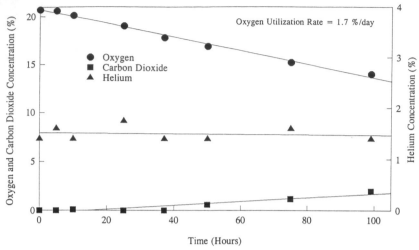

Figure 5-6. In Situ Respiration Test Results with Linear Oxygen Concentration Versus Time at AOC A, Keesler AFB, Mississippi

Table 5-7
Oxygen Utilization and Carbon Dioxide Production Rates During the In Situ Respiration Test at AOC A, Keesler AFB, Mississippi

Sample Name	Oxygen Utilization Rate (%/hour)	Estimated Biodegradation Rate (mg/kg-day)
K1-MPA-5.0′	0.071	1.16
K1-MPA-7.0′	0.045	0.73
K1-MPB-5.0′	0.13	2.12
K1-MPC-7.0	0.099	1.62
Background	0.012	0.20

Table 5-8
Raw Data From an In Situ Respiration Test at SWMU 66, Keesler AFB, Mississippi

Time (Hours)	Oxygen (%)	Carbon Dioxide (%)	Helium (%)
0	20.5	0	1.6
6.3	18.1	.05	1.6
9.3	16.5	1.0	1.6
15	14	2.2	1.8
22	11	3.2	1.5
31	6.8	5.0	1.5
48	3.7	5.1	1.5
57	2.9	5.1	1.5

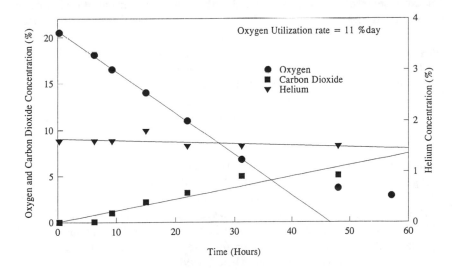

Figure 5-7. In Situ Respiration Test Results with Nonlinear Oxygen Concentration
Versus Time at SWMU 66, Keesler AFB, Mississippi

The helium data collected at a site will provide insight into whether observed oxygen utilization rates are due to microbial utilization or to other effects such as leakage or diffusion. As a rough estimate, diffusion of gas molecules is inversely proportional to the square root of the molecular weight of the gas. Based on the molecular weights of 4 and 32 g/mole for helium and oxygen, respectively, helium diffuses about 2.8 times faster than oxygen. Thus, although helium is a conservative tracer, its concentration should decrease with time. As a general rule of thumb, one should consider any in situ respiration test in which the rate of helium loss is less than the oxygen loss rate to be an acceptable test. If the helium loss rate is greater than the oxygen loss rate, disregard the test from that monitoring point. We do not use the helium loss rate to correct the oxygen utilization rate.

Example 5-6. *Evaluation of Helium Loss During an In Situ Respiration Test:* Figures 5-8 and 5-9 show helium data for two test wells. The helium concentration at monitoring point S1 (Figure 5-8) at Tinker AFB started at 1.5% and after 108 hours had dropped to 1.1%, i.e., a fractional loss of ~0.25; therefore, S1 was an acceptable point. In contrast, for Kenai K3 (Figure 5-9), the change in helium was rapid (a fractional drop of about 0.8 in 7 hours), indicating that there was possible short circuiting at this monitoring point. This suggested that the data from this monitoring point were unreliable, and the data were not used in calculating degradation rates.

C. FACTORS AFFECTING OBSERVED IN SITU BIODEGRADATION RATES

Because in situ biodegradation rates are measured indirectly through measurements of soil gas oxygen and carbon dioxide concentrations, abiotic processes that affect oxygen and carbon dioxide concentration will affect measured biodegradation rates. The factors that may most influence soil gas oxygen and carbon dioxide concentrations are soil pH, soil alkalinity, and iron content. In addition, any environmental parameter that may affect microbial activity also may affect observed oxygen utilization rates. Soil temperature often is a significant factor at bioventing sites.

At several sites, oxygen utilization has proven to be a more useful measure of biodegradation rates than carbon dioxide production. The biodegradation rate in mg of hexane-equivalent/kg of soil per day based on carbon dioxide production usually is less than can be accounted for by the oxygen disappearance. At virtually all sites studied as part of the U.S. Air Force Bioventing Initiative, oxygen utilization rates have been higher than carbon dioxide production rates. However, a study conducted at Tyndall AFB site was an exception. That site had low-alkalinity soils and low-pH quartz sands, and carbon dioxide production actually resulted in a slightly higher estimate of biodegradation (Miller, 1990).

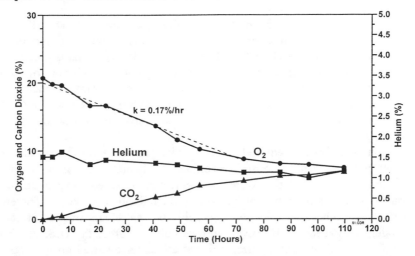

**Figure 5-8. In Situ Respiration Test Results With Acceptable Data Based on the
Helium Concentration for Monitoring Point S1, Tinker AFB, Oklahoma**

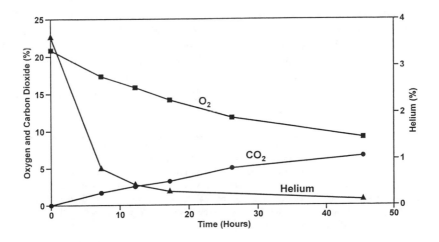

**Figure 5-9. In Situ Respiration Test Results With Unacceptable Data Based on the
Helium Concentration for Monitoring Point K3, Kenai, Alaska**

In the case of the higher pH and higher alkalinity soils at Fallon NAS and Eielson AFB, little or no gaseous carbon dioxide production was measured (Hinchee et al., 1991a; Leeson et al., 1995). This is possibly due to the formation of carbonates from the gaseous evolution of carbon dioxide produced by biodegradation at these sites. A similar phenomenon was encountered by van Eyk and Vreeken (1988) in their attempt to use carbon dioxide evolution to quantify biodegradation associated with soil venting.

Iron is a nutrient required for microbial growth, but the iron also may react with oxygen to form iron oxides. Theoretically, if a significant amount of iron oxidation were to occur, the observed oxygen utilization rate would reflect both iron oxidation and microbial activity. Therefore, calculated biodegradation rates would be an overestimate of actual biodegradation rates. However, in data collected from the U.S. Air Force Bioventing Initiative study, iron concentrations have varied greatly, ranging from less than 100 mg/kg to greater than 100,000 mg/kg, with no apparent impact on oxygen utilization rates. Iron impact on oxygen utilization rates has been observed at only one site, the Marine Base at Kaneohe, Hawaii, where soil iron concentrations are in the 100,000 mg/kg range.

It is important to consider whether the respiration rate was measured at the time of year when microbial activity rates were at their maximum (summer) or if it was measured when activity was low (winter). Investigations at a number of sites have shown that microbial rates can vary by as much as an order of magnitude between peak periods. For design of oxygen delivery systems, respiration rates should be measured during the peak season, typically late summer.

If oxygen utilization rates were determined during periods of low activity, it will be necessary to adjust the rates to the maximum level before making size calculations. The van't Hoff-Arrhenius equation can be used to predict oxygen utilization rates given an initial rate and temperature[1]. The activation energy, E_a must either be known for the site or calculated by using E_a found at another site, recognizing that the temperature-adjusted rate is only a rough estimate. The following example illustrates a typical adjustment.

Example 5-7. Temperature Adjustment of Oxygen Utilization Rate: The oxygen utilization rate was measured in January at a site in Cheyenne, Wyoming. The rate was determined to be 0.75%/day (0.031%/hr). The temperature in the soil was measured at 4°C. Previous temperature measurements at the site have indicated that soil temperatures in August average approximately 24°C, i.e., 20°C higher than the temperature measured during January. The temperature adjustment to the rate for sizing calculations is as follows:

---

[1] Refer to Chapter 3 for a discussion of the effect of temperature on microbial activity.

Using the van't Hoff-Arrhenius equation (Metcalf & Eddy, 1979):

$$\frac{dk}{dT} = \frac{E_a}{RT^2}$$

Integration of this equation between the limits T_1 (277°K) and T_2 (297°K) gives:

$$\ln \frac{k_T}{k_o} = \frac{E_a (T_2 - T_1)}{RT_1T_2}$$

where: k_T = temperature-corrected oxygen utilization rate (% O_2/day)
k_o = baseline reaction rate = 0.75%/day
E_a = activation energy[1] = 13.4 kcal/mole
R   = gas constant = 1.987 cal/°K-mole
T_1 = absolute temperature for k_o = 277°K
T_2 = absolute temperature for k_T = 297°K

$$k_T = \left(0.75\ \frac{\%}{day}\right) e^{\left[\frac{(13,400\,cal/mole)(297°K\, -\, 277°K)}{\left(1.987\frac{cal}{°K\,mole}\right)(297°K)(277°K)}\right]}$$

$$k_T = 3.9\frac{\%}{day}$$

As can be seen from this calculation, the site would require approximately 5 times greater oxygen delivery rate in the summer.

IV. SOIL GAS PERMEABILITY AND RADIUS OF INFLUENCE

In situ respiration rates may be used to calculate the required air flowrate to satisfy oxygen demand at a given site[2]. However, it is necessary also to determine the distance air can physically be moved. An estimate of the soil's permeability to fluid flow (k) and the radius of influence (R_I) of venting wells are both important elements of a full-scale bioventing design. On-site testing provides the most accurate estimate of the soil gas permeability. On-site testing also can be used to determine the radius of influence that can be achieved for a given well configuration and flowrate. These data are used in full-scale system design, to space venting wells, to size blower equipment, and to ensure that the entire site receives a supply of oxygen-rich air to sustain in situ biodegradation.

Soil gas permeability, or intrinsic permeability, can be defined as a soil's capacity for fluid flow, and varies according to grain size, soil

---

[1] Calculated from a different field site. Refer to Example 3-2 for a description of the calculation of the activation energy.

[2] Refer to Chapter 6 for a presentation of the calculation of required air flowrates.

uniformity, porosity, and moisture content. The value of k is a physical property of the soil; k does not change with different extraction/injection rates or different pressure levels.

Soil gas permeability is generally expressed in the units cm^2 or darcy (1 darcy = 1 x 10^{-8} cm^2). Like hydraulic conductivity, soil gas permeability may vary by more than an order of magnitude at one site because of soil variability. Table 5-9 illustrates the range of typical k values to be expected with different uniform soil types. Actual soils will contain a mixture of grain sizes, which generally will increase the observed darcy values based on pilot testing.

Several field methods have been developed for determining soil gas permeability (Sellers and Fan, 1991). The most commonly applied field test method probably is the modified field drawdown method developed by Paul Johnson at Arizona State University and former associates at the Shell Development Company. This method involves the injection or extraction of air at a constant rate from a single venting well while measuring the pressure/vacuum changes over time at several monitoring points in the soil away from the venting well.[1]

The field drawdown method is based on Darcy's law and equations for steady-state radial flow to or from a vent well. A full mathematical development of this method and supporting calculations are provided by Johnson et al. (1990). The HyperVentilate™ computer program was produced by Johnson for use in storing field data and computing soil gas permeability. This or other commercially available programs can be used to speed the calculation and data presentation process.

Two solution methods may be used to calculate soil gas permeability, as described in Johnson et al. (1990). The first solution is based on carefully measuring the dynamic response of the soil to a constant injection or extraction rate. The second solution for soil gas permeability is based on steady-state conditions and the measurement or estimation of the radius of influence at steady state. Whenever possible, field data should be collected to support both solution methods because one or both of the solution methods may be appropriate, depending on site-specific conditions. An example procedure for conducting a soil gas permeability test is provided in Appendix C.

---

[1] Refer to Appendix B for recommended specifications and manufacturers for the soil gas permeability testing equipment.

Table 5-9
Soil Gas Permeability Values (Johnson et al., 1990)

Soil Type	k in Darcy
Coarse sand	100 to 1,000
Medium sand	1 to 100
Fine sand	0.1 to 1.0
Silts/clay	<0.1

A. RADIUS OF INFLUENCE DETERMINATION BASED ON PRESSURE MEASUREMENTS

At a bioventing site, the radius of influence is defined as the maximum distance from the air extraction or injection well where a sufficient supply of oxygen for microbial respiration can be delivered. We will call the radius of influence measured by increased oxygen the "oxygen radius of influence". In practice, we frequently estimate this radius by measuring a pressure radius of influence. A description of how that is done will follow.

The oxygen and pressure radii of influence are a function of soil properties, but also are dependent on the configuration of the venting well and extraction or injection flowrates, and are altered by soil stratification. The oxygen radius of influence also depends on microbial oxygen utilization rates. At sites with shallow contamination, the oxygen and pressure radius of influence also may be increased by impermeable surface barriers such as asphalt or concrete. These paved surfaces may or may not act as vapor barriers. Without a tight seal to the native soil surface[1], the pavement will not significantly impact soil gas flow.

At a bioventing site, the oxygen radius of influence is the true radius of influence; however, for design purposes, we frequently use the pressure radius of influence. The pressure radius of influence is the maximum distance from a vent well where vacuum (in extraction mode) or pressure (in injection mode) can be measured. In practice, we usually use 0.1 inches of water as the cut off pressure. In highly permeable soils, 0.01 inches of water is a better cut off, if it can be reliably measured. There is a connection between the pressure radius of influence and the oxygen radius of influence; however, there are many variables which are not fully understood. In practice, it has been our experience that when our design procedures are followed, that the oxygen radius of influence is larger than the measured pressure radius of influence, making the pressure radius of

---

[1] It is the authors' experience that at most sites, this seal does not occur.

96

influence a reasonably conservative, rapid method for estimating the true
radius of influence. The oxygen radius of influence may be determined
directly by measuring the distance from the vent well at which a change in
oxygen concentration can be detected. However, it may take several days
to weeks for equilibrium to be reached and an accurate oxygen radius of
influence to be measured. In addition, oxygen utilization rates may
change, increasing or reducing the oxygen radius of influence. Therefore,
if possible, it is best to measure the oxygen radius of influence at times of
peak microbial activity. Alternatively, the pressure radius of influence may
be determined very quickly, generally within 2 to 4 hours. Therefore, the
pressure radius of influence typically is used to design bioventing systems.

The pressure radius of influence should be determined at three different
flowrates, with a 1- to 2-hour test per flowrate during the permeability test.
Determining the radius of influence at different flowrates will allow for
more accurate blower sizing[1]. Recommended flowrates for the
permeability test are 0.5, 1.5, and 3 cfm (14, 42, 85 L/min) per ft (0.3 m)
of well screen.

The pressure radius of influence may be estimated by determining
pressure change versus distance from the vent well. The log of the
pressure is plotted versus the distance from the vent well. The radius of
influence is that distance at which the curve intersects a pressure of
0.1"H_2O (25 Pa). This value was determined empirically from U.S. Air
Force Bioventing Initiative sites. Example 5-8 illustrates calculating the
radius of influence in this manner.

*Example 5-8. Calculation of the Radius of Influence Based on Pressure
Measurements:* Soil gas permeability results from the Saddle Tank Farm Site at Galena
AFS, Alaska are shown in Figure 5-10 with the log of the steady-state pressure
response at each monitoring point plotted versus the distance from the vent well. The
radius of influence is taken to be the intersection of the resulting slope of the curve at a
pressure of 0.1"H_2O (25 Pa). Therefore, in this instance, the pressure radius of
influence would be estimated at 92 ft (28 m).

When using pressure radius of influence, it should be remembered that
the estimated radius of influence actually is an estimate of the radius in
which measurable soil gas pressures are affected and does not always
equate to gas flow. In highly permeable gravel, for example, significant
gas flow can occur well beyond the measurable radius of influence. On the
other hand, in a low-permeability clay, a small pressure gradient may not
result in significant gas flow.

[1] Refer to Chapter 6 for a discussion of blower sizing.

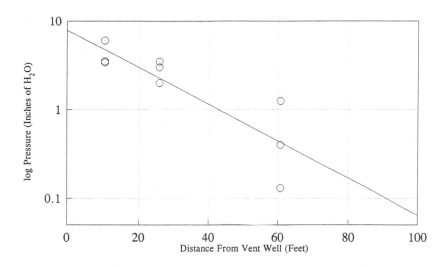

Figure 5-10. Determination of Radius of Influence at the Saddle Tank Farm, Galena AFS, Alaska

B. INTERPRETATION OF SOIL GAS PERMEABILITY TESTING RESULTS

The technology of bioventing has not advanced far enough to provide firm quantitative criteria for determining the applicability of bioventing based solely on values of soil permeability or the radius of influence. In general, the soil permeability must be sufficiently high to allow movement of oxygen in a reasonable time frame (1 to 10 days) from either the vent well, in the case of injection, or the atmosphere or uncontaminated soils, in the case of extraction. If such a flowrate cannot be achieved, oxygen cannot be supplied at a rate to match its demand. Closer vent well spacing or high injection/extraction rates may be required. If either the soil gas permeability or the radius of influence is high (>0.01 darcy or a R_I greater than the screened interval of the vent well), this is a good indicator that bioventing may be feasible at the site and it is appropriate to proceed to soil sampling and full-scale design. If either the soil gas permeability or the radius of influence is low (<0.01 darcy or a R_I less than the screened interval of the vent well), bioventing may not be feasible. In this situation, it is necessary to evaluate the cost effectiveness of bioventing over other alternative technologies for site remediation.   The cost of installing a bioventing system at a low-permeability site will be driven primarily by the need to install more vent wells, use a blower with a higher delivery pressure, or install horizontal wells.

98

V. SOIL CHARACTERIZATION

Soil characterization is a critical component of the site characterization process. Of primary importance is the concentration and distribution of contaminants. Because there typically are large variations in the distribution of contaminants at a site, a relatively large number of soil samples must be collected to adequately delineate the vertical and areal extent of contamination. Described in the following sections are techniques for locating and drilling soil borings[1]. The soil analytical protocol is also discussed.

A. SOIL BORINGS

Soil borings should be located based on either the review of existing site data or the results of the soil gas survey. Soil borings can serve two purposes: the collection of soil samples and the installation of vent wells and monitoring points. Soil borings have the advantage of allowing a large number of soil samples to be collected from a single location and allowing for subsequent installation of the vent wells and monitoring points in the borings. Disadvantages include the generation of soil cuttings and the fact that drilling may require subcontracting and a large amount of time. Alternative methods, such as a GeoProbe™ system or cone penetrometer, may be used for collection of soil samples and may be suitable for installing soil gas monitoring points.

The hollow-stem auger method is generally preferred for drilling in unconsolidated soils; however, a solid-stem auger is acceptable in more cohesive soils. The final diameter of the borehole is dependent on the diameter selected for the vent wells, but typically should be at least two times greater than the outside diameter of the vent well.

All drilling and sample collection activities should be observed and recorded on a geologic boring log. Data to be recorded includes soil sample interval, sample recovery, visual presence or absence of contamination, soil description, and lithology. Soil samples should be labeled and properly stored immediately after collection. An example procedure for soil sample collection, labeling, packing, and shipping is provided in Appendix C.

It is preferable that all boreholes be completed as vent wells or monitoring points. If this is not possible, boreholes must be abandoned according to applicable state or federal regulations. Typically, borehole abandonment is accomplished by backfilling with bentonite or grout.

---

[1] Refer to Appendix B for recommended specifications and manufacturers for the soil sampling equipment.

B. SOIL ANALYSES

A summary of soil analyses is provided in Table 5-10. Methods in this table are not the only methods available, but are those currently used by the Air Force.

Results of the U.S. Air Force Bioventing Initiative indicate that four parameters should always be measured: aromatic hydrocarbons (BTEX), total petroleum hydrocarbons (TPH), moisture content, and particle size. Another measurement, total Kjeldahl nitrogen (TKN), was found to be a statistically significant factor in the statistical analyses of U.S. Air Force Bioventing Initiative data[1]; but there is no evidence to date that addition of nitrogen will enhance site remediation.

Measurements of BTEX and TPH are necessary for delineation of the contaminant plume. In addition, BTEX and TPH typically are of regulatory concern; therefore, these concentrations must be established. Moisture content has been found to limit biodegradation in extreme environments. At a site in California, moisture content averaged approximately 2% and irrigation did substantially improve biodegradation rates[2]. Particle size distribution is an important indicator of permeability. High clay content soils may be difficult to biovent due to the inability to move air through the soil particularly when high moisture levels exist. In addition, clay particles can be sites of significant contaminant adsorption and as such can significantly affect contaminant sorption and bioavailability.

TKN is a nutrient required for microbial growth and respiration; therefore, low TKN levels may affect microbial respiration. However, while a statistically significant relationship has been observed between TKN and oxygen utilization rates, the relationship is weak and unlikely to have practical significance. Therefore, it is only recommended to analyze for TKN if all other explanations for poor bioventing performance have been exhausted (i.e., permeability, moisture content).

---

[1] Refer to Section 5.0, Volume I for a discussion of the statistical analyses of Bioventing Initiative data.

[2] Refer to Section 3.2.2.2, Volume I for a discussion of this site.

100

Table 5-10
Soil Analyses Based on U.S. Air Force Bioventing Initiative Results

Analysis[1]	Method	Comments	Sample Volume, Container, Preservation	Field or Analytical Laboratory
BTEX	Purge and trap GC method SW8020	Handbook method	Collect 100 g of soil in brass sleeves[2]; store at 4°C until analyzed.	Analytical laboratory
TPH	Modified GC method SW8015	Handbook method; reference CA LUFT[3] manual	Collect 100 g of soil in brass sleeves[2]; store at 4°C until analyzed.	Analytical laboratory
Moisture Content	ASTM D-2216	Handbook method	Collect in a 4-oz glass container with Teflon™-lined cap.	Analytical laboratory
Particle Size Analysis	ASTM D422	Handbook method	Collect 250 g soil in a glass or plastic container.	Analytical laboratory
TKN[4]	EPA 351.4	Handbook method	Collect in a 4-oz glass container with Teflon™-lined cap.	Analytical laboratory

[1] Recommended soil analyses are based on experience and analyses of petroleum-contaminated sites. Additional data may be required at sites contaminated with other compounds. [2] One sample in the brass sleeves provides sufficient volume for analyses of both aromatic hydrocarbons and TPH. [3] LUFT = State of California *Leaking Underground Fuel Tank Field Manual*, 1988 edition. [4] Not recommended for an initial analysis, but only if bioventing performance is poor and other factors such as permeability and moisture content do not account for the poor performance.

CHAPTER 6

BIOVENTING IMPLEMENTATION:
SYSTEM DESIGN

The design of a bioventing system is based on the results of site characterization and pilot testing efforts described in Chapter 5. The objective is to design a system that results in aeration of the contaminated soils with little or no volatilization. Aeration may be accomplished through air injection, gas extraction, or a combination of the two. Soil vacuum extraction (a.k.a. soil venting, soil gas extraction, or vacuum vapor extraction) is a related technology in which soil gas is extracted to remove contaminants by volatilization. In contrast, bioventing is designed to minimize volatilization and optimize biodegradation. As a result, bioventing typically uses much lower air flowrates and often does not involve air extraction.

The basic steps involved in designing a bioventing system are described in this chapter as follows:

1. Determine required air flow system (injection, extraction, or both).
2. Determine required air flowrates.
3. Determine the working radius of influence.
4. Determine well spacing.
5. Provide detailed design of blower, vent wells, and piping.
6. Determine vent well requirements.
7. Determine monitoring point requirements.

I. DETERMINATION OF AIR FLOW SYSTEM

In general, if safe and feasible, air injection is the preferred configuration for full-scale bioventing systems. If properly designed, air injection will result in minimal discharge of volatile organics to the atmosphere and is less expensive to operate and maintain than air extraction systems.

Under some circumstances, soil gas extraction systems may need to be incorporated into an air injection system design. For example, whenever the radius of influence of a vent well reaches basements, utility corridors, or occupied surface structures, an air extraction system will reduce the risk of moving gases into these areas. This precaution will prevent the accumulation of explosive or toxic vapors in these structures.

101

A. AIR INJECTION

Air injection involves the introduction of air under pressure into the contaminated zone. If the contaminants are volatile, some will migrate in the gas phase into surrounding soil where they can biodegrade. This has the advantage of creating an expanded in situ bioreactor as illustrated in Figure 6-1. Given adequate oxygen, the volatilized hydrocarbons will biodegrade in these surrounding uncontaminated soils, increasing the fraction of contaminants biodegraded compared to an air extraction configuration. This concept is illustrated in Example 6-1.

Example 6-1. Biodegradation of Petroleum Hydrocarbons in the Uncontaminated and Contaminated Regions at Site 280, Hill AFB: At this site, high vapor phase TPH concentrations were detected within a radius of approximately 50 ft (15 m) from the injection well. TPH concentrations decreased with increasing distance from the well. Likewise, in situ respiration rates were observed to decrease with increasing distance from the injection well (Figure 6-2). Calculations were made to compare total TPH mass degraded in each region based on these in situ respiration rates. These results, shown in Figure 6-3, illustrate that, despite relatively low in situ respiration rates at monitoring points located far from the injection well (220 ft [67 m]), the majority of the contaminant degradation was occurring in this area. These results illustrate the availability of vapor-phase hydrocarbons for biodegradation and the significant contribution an expanded bioreactor can have on contaminant removal.

Miller (1990) found at the Tyndall AFB site that hydrocarbon vapors biodegrade at approximately one-third the rate observed in contaminated soils. Kampbell (1993) found that vapor phase biodegradation in an air injection system was greatest in shallow root zone soils. The concept is analogous to an in situ biofilter. In general, air can be injected at flowrates low enough to avoid surface emissions. As air injection rate increases, hydrocarbon volatilization increases (Figure 6-4). Therefore, the objective is to inject sufficient air to meet oxygen demand for biodegradation but not to cause emissions to the atmosphere. This is generally possible at sites contaminated with JP-4 or JP-5 jet fuel, diesel, and other contaminants of similar or lesser volatility. It is more difficult with gasoline, although successful systems using only air injection have been reported at gasoline-contaminated sites (Kampbell, 1993; Reisinger, 1994).

In addition to creating an expanded bioreactor, air injection has the potential to expose a significant portion of capillary fringe contaminated soil to treatment via water table depression. As air is injected into the vadose zone, a positive pressure is created, resulting in depression of the water table. Figure 6-5 illustrates the water table depression observed at Site 20, Eielson AFB, Alaska. This water table depression has important

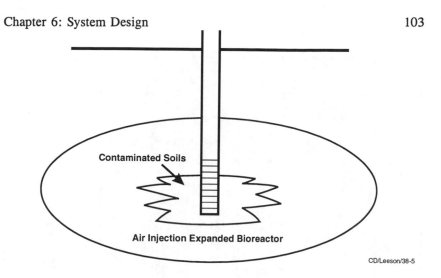

Figure 6-1. Expanded Bioreactor Created During Air Injection

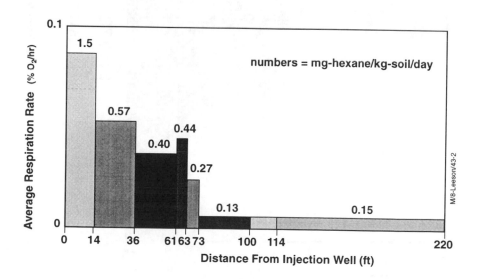

Figure 6-2. Oxygen Utilization Rates, Averaged Over Depth, Versus Distance from the Injection Well at Site 280, Hill AFB, Utah

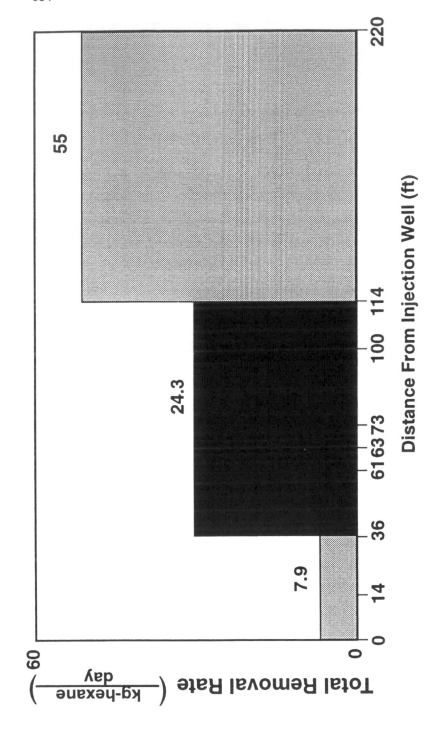

Figure 6-3.   Mass of TPH Degraded Versus Distance from the Injection Well at Site 280, Hill AFB, Utah

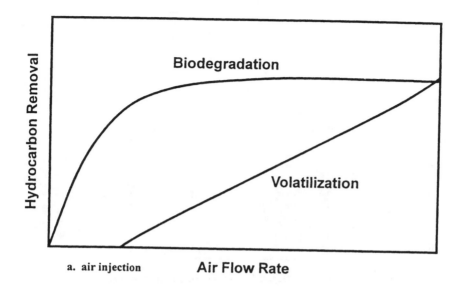

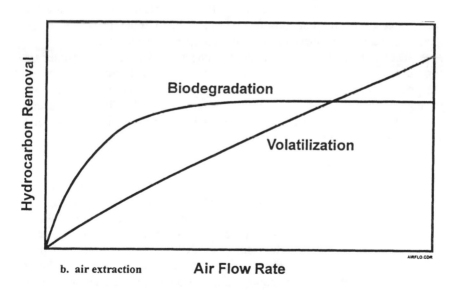

Figure 6-4. Hydrocarbon Volatilization and Biodegradation Rates as a Function of Air Flowrate

106

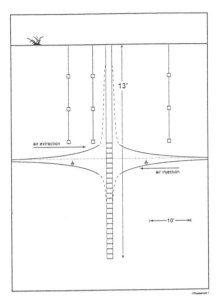

Figure 6-5. Water Table Depression During Air Injection and Air Extraction

implications. At many sites, the capillary fringe is highly contaminated, and the capillary fringe will be more effectively treated by lowering the water table. In addition, this dewatering effect frequently results in an increased radius of influence and greater soil gas permeability.

A schematic diagram of a basic air injection system is illustrated in Figure 6-6. The system is relatively simple, involving a blower or compressor and a distribution system. Explosion-proof blowers are recommended for safety. If properly designed and operated, an injection system will not result in significant air emissions or require aboveground vapor phase treatment.

B. AIR EXTRACTION

Air injection is the preferred bioventing configuration; however, air extraction may be necessary at sites where movement of vapors into subsurface structures or air emissions are difficult to control. If a building or other structure is located within the radius of influence of a site, or if the site is near a property boundary beyond which hydrocarbon vapors cannot be pushed, air extraction may be considered. A significant disadvantage of the air extraction configuration is that biodegradation is limited to the contaminated soil volume because vapors do not move outward to create an expanded bioreactor. The result is less biodegradation and more volatilization. In general, increasing extraction rates will increase both volatilization and biodegradation rates until the site becomes

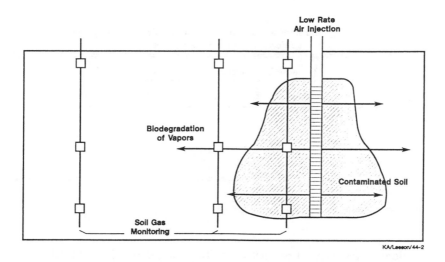

Figure 6-6. Air Injection Configuration for a Bioventing System

aerated, above which the rate of biodegradation no longer increases. Volatilization generally will continue to increase with increasing extraction rates until the contaminated soil system becomes diffusion-limited. The optimal air flowrate for both injection and extraction is the minimum required to satisfy the oxygen demand. Extraction systems result in some volatilization regardless of the extraction rate. Figure 6-4 illustrates this concept. The relative removal attributed to biodegradation and volatilization is quite variable and site-dependent. At a JP-4 jet fuel-contaminated site at Tyndall AFB, Miller et al. (1991) found that at the optimal air injection level it was possible to achieve approximately 85% of removal due to biodegradation at the optimal flowrate.

Air extraction creates a partial vacuum in the soil, resulting in a water table and capillary fringe rise or upwelling. This phenomenon has been illustrated in the soil venting literature (Johnson et al., 1990). Because the bulk of contamination often lies several inches or feet above or below the water table (smear zone), this upconing can saturate much of the contaminated soil and reduce treatment efficiency. The upconing also will increase soil moisture in the capillary fringe and thus reduce soil gas permeability and radius of influence.

An example of this phenomenon was observed at Eielson AFB. An extraction air permeability test was conducted at Eielson AFB to observe the effect of the bioventing configuration on the site air permeability and well radius of influence. Table 6-1 compares the results of extraction and

Table 6-1
Permeability and Radius of Influence Values at Eielson AFB, Alaska:
Injection and Extraction Mode

Depth (ft)	Permeability (darcy)		Air Radius of Influence (ft)	
	Injection	Extraction	Injection	Extraction
2	NR	NR	<7.0	<6.0
4	0.53	0.27	45	34
6	0.56	0.27	68	42

NR = No response

injection tests at Site 20 on Eielson AFB. The permeability (k) calculated for the extraction test was 0.27 darcy, approximately one-half the result for the air injection test. The radius of influence observed at the 6-ft (1.8-m) monitoring depth also was reduced approximately one-third to 42 ft (13 m) (Figure 6-7). This reduction in permeability and radius of influence was a result of the water table rise illustrated in Figure 6-5.

Figure 6-8 is a schematic of a basic air extraction system. In contrast to an injection system, an explosion-proof blower with explosion-proof wiring normally is required. Extracted soil gas typically contains moisture at or near saturation, and a knockout (air/water separator) usually is required to collect condensate, which must be treated or disposed of. Also, in winter months in regions with sustained temperatures below freezing, insulation and/or heat tape may be required to maintain piping at temperatures above freezing to avoid clogged pipes.

Air extraction systems usually will result in point source emissions that may require permitting and treatment. Air treatment will increase remediation costs significantly. Appendix D contains information on options for off-gas treatment.

Currently, air extraction has been selected as the method for oxygenation for only 6 out of the 125 U.S. Air Force Bioventing Initiative sites. Two of the sites (Capehart Service Station at McClellan AFB and BX Service Station, Patrick AFB) operated in extraction mode for 60 to 120 days, at which time the system was reconfigured for air injection because vapor concentrations had been significantly reduced. At Patrick AFB, initial vapor concentrations of TPH were as high as 27,000 ppmv (Figure 6-9). After approximately 75 days of operation, concentrations were reduced to 1,600 ppmv and the bioventing system was reconfigured for injection

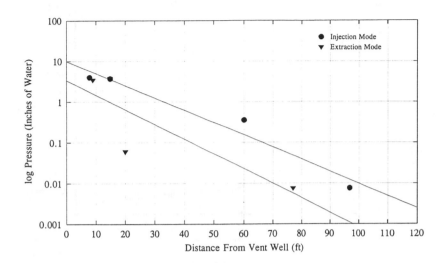

Figure 6-7. Radii of Influence During Air Injection and Extraction in the Control Test Plot at a Depth of 6 ft at Site 20, Eielson AFB, Alaska

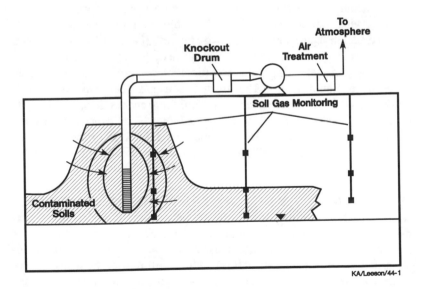

Figure 6-8. Schematic Diagram of a Basic Air Extraction System

110

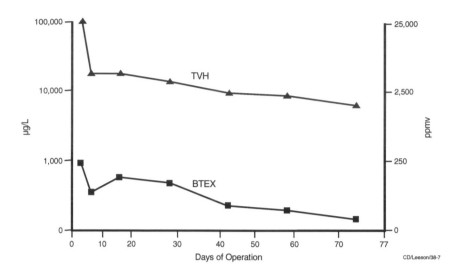

Figure 6-9. Extracted BTEX and TPH Soil Gas Concentrations at Patrick AFB, Florida

(Engineering-Science, 1994b).   The Base Service Station at Vandenberg AFB contained high concentrations of more volatile components of gasoline and is an active service station. As such, the possibility of vapor migration into the service station was possible. This bioventing system was operated in an extraction configuration in two Phases (Downey et al., 1994a). During Phase I, extracted soil gas was passed through a PADRE® vapor treatment system, where high concentrations of volatiles were adsorbed and condensed to liquid fuel.   The treated soil gas then was recirculated through the soil by injecting air via biofilter trenches located along the perimeter of the site. Phase II was initiated once TVH concentrations were reduced to <1,000 ppmv. At this time, the PADRE® system was taken off line, and the extracted soil gas was reinjected directly into the biofilter trenches.

C. DETERMINING USE OF INJECTION VERSUS EXTRACTION
The decision to use injection versus extraction usually is driven by safety considerations.   Air injection should not be used unless a system can be designed that will not push hazardous vapors into structures.   Table 6-2 summarizes some of the considerations which will impact the decision.

Numerous options are available that may allow air injection at sites with structures at risk or with property boundaries nearby (Downey et al., 1995). These options include monitoring the atmosphere in the structure to verify that no contaminant has entered, using air extraction coupled with

Table 6-2
Air Injection Versus Extraction Considerations

Favor Injection	Favor Extraction
Low vapor pressure contaminants	High vapor pressure contaminants
Deep contamination	Surface emissions concern
Low permeability soils	Structures/property boundaries within the radius of influence
Significant distance from structures/property boundaries	

reinjection to protect the building (Figure 6-10), or using subslab depressurization.

D. DESIGN OF AIR FLOW TO PROTECT STRUCTURES

Subslab depressurization can be used to protect structures while still allowing for air injection to provide optimal oxygenation. Subslab depressurization involves extracting air within or around the perimeter of a building during simultaneous air injection. Vapors extracted from beneath the building may be released to the atmosphere, treated then released, or reinjected into the subsurface for further biotreatment. A schematic diagram of such a system is shown in Figure 6-11.

At AOC A at Keesler AFB, Mississippi, a subslab depressurization system is currently in operation as part of the U.S. Air Force Bioventing Initiative. A schematic diagram of the site is shown in Figure 6-12. Soil vapor is continually withdrawn from air extraction wells located around the perimeter of the building and reinjected into the vent wells. Makeup air is added to the injection gas to provide sufficient oxygen to aerate the site. No vapor migration into the building has been detected at this site, and the site soils are well-oxygenated.

At Site 48 at Eielson AFB, Alaska, a utilidor that is actively used runs through the site. The potential for migration of vapors into the utilidor was high. To eliminate vapor migration into this structure, a horizontal perforated pipe was installed next to the utilidor. A vertical extraction well was connected to the horizontal pipe to extract gas from along the utilidor for vapor control. The extracted soil gas then was reinjected into a contaminated area at the site (Figure 6-13).

112

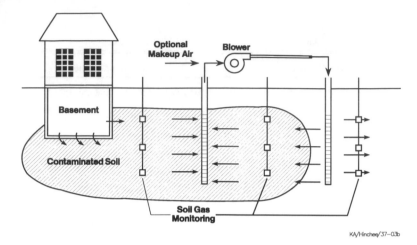

KA/Hinchee/37-03b

Figure 6-10. Schematic Diagram of an Air Injection System with Reinjection of Extracted Soil Gas

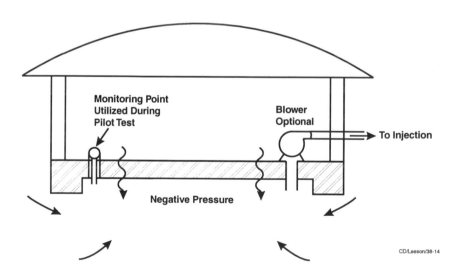

CD/Leeson/38-14

Figure 6-11. Schematic Diagram of Subslab Depressurization

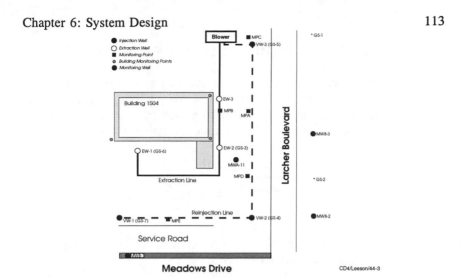

Figure 6-12. Schematic Diagram of the Extraction with Reinjection System at AOC A, Keesler AFB, Mississippi

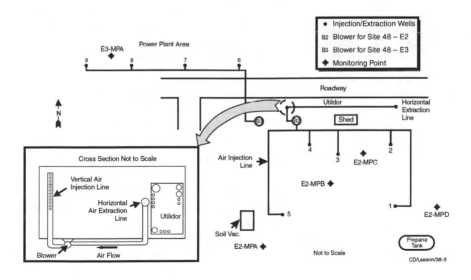

Figure 6-13. Soil Gas Extraction to Isolate a Subsurface Structure at Site 48, Eielson AFB, Alaska

II. DETERMINING REQUIRED AIR FLOWRATES

The flowrate required to operate the bioventing system is dependent on the oxygen demand of the indigenous microorganisms. This is best determined from maximum oxygen utilization rates measured during an in situ respiration test. Equation 2-1 is used to estimate the required air flowrate:

$$Q = \frac{k_o \, V \, \theta_a}{(20.9\% - 5\%) \times 60\frac{min}{hr}}$$

(6-1)

where: Q = flowrate (ft^3/min)
k_o = oxygen utilization rate (%/hr)
V = volume of contaminated soil (ft^3)
θ_a = gas-filled porosity (fraction, i.e. 0.2 or 0.3)

Example 6-2 illustrates the use of this calculation.

Example 6-2. *Determination of Required Air Flowrate:* Given a volume of contaminated soil of approximately 170,000 ft^3 (4,760 m^3), an air-filled void volume (θ_a) at this site of 0.36[1], and an oxygen utilization rate of 0.25%/hr, the flowrate is calculated as follows:

$$Q = \frac{(0.25 \ \%/hr)(170,000 \ ft^3)(0.36)}{(20.9\% - 5\%) \times 60 \ min/hr}$$

Therefore, the required flowrate is approximately 16 cfm (453 L/min).

The flowrate selected from this calculation must be confirmed during bioventing system operation by monitoring soil gas composition to ensure adequate oxygen levels at all locations. The calculated flowrate also can be compared to flowrates used for the soil gas permeability test to determine whether the calculated flowrate will provide sufficient aeration. Higher flow rates than that calculated by this method may be necessary for adequate site aeration.

Data from numerous sites contaminated with various types and mixtures of contaminants have shown that microbial activity is not oxygen-limited above oxygen concentrations of approximately 1 to 2%. To ensure adequate oxygen levels in the entire treatment cell, a minimum level of 5% should be maintained.

---

[1] Refer to Chapter 5 on using moisture content to estimate this parameter.

III. WELL SPACING

To determine the required number of wells and the appropriate spacing, an estimate of the radius of influence is necessary.   A number of approaches to this are possible. Those normally in use are:

- Based on measured pressure in monitoring points during a soil gas permeability test.
- Estimated from air flow and oxygen consumption.
- Measured empirically.

Estimating the radius of influence based on pressure measurements during an in situ permeability test is a common approach used in soil venting or soil vapor extraction and probably is the fastest method. It normally is done by plotting the log of pressure versus distance as described in Section 1.5.3. The limitation to this approach is that it incorporates only one of the three factors that affect the radius of influence. In order to determine more exactly the radius of oxygen influence, air flowrate and oxygen utilization must be considered. In low-permeability soils, a pressure effect may be seen in a monitoring point, but air flowrates to that point may be too low to supply adequate oxygen. Conversely, in a high-permeability soil, air flowrates sufficient to supply oxygen may occur at pressure differentials that cannot be measured.   It has been our experience that, if a pressure criteria of $0.1''II_2O$ (25 Pa) is used, the estimated radius of influence will be conservative for well spacing and site aeration.

Radius of influence for a given air flowrate can be estimated based on oxygen utilization. Assuming the use of a vertical well so that air flow can be described in cylindrical coordinates and assuming that the radius of influence is much greater than the well radius, the following equation can be used:

$$R_I = \sqrt{\frac{Q(20.9\% - 5\%)}{\pi \, h \, k_o \, \theta_a}} \qquad (6\text{-}2)$$

where: R_I = radius of influence (ft)
$\quad\quad\;\; Q$ = air flowrate (ft^3/day)
$\quad\quad\;\;$ 20.9 - 5% = oxygen %
$\quad\quad\;\; k_o$ = oxygen utilization rate (%/day)
$\quad\quad\;\; \theta a$ = air filled porosity (cm$^3{}_{air}$/cm$^3{}_{soil}$)
$\quad\quad\;\; h$ = aerated thickness (ft)

116

Example 6-3. Calculation of Radius of Influence: To calculate the radius of influence at Dover AFB, Equation 6-2 is used with the following parameters:

Q = 20 cfm (570 L/min) = 28,800 ft³/day (820,800 L/day)
k_o = 4%/day
θ_a = 0.25
h = 20 ft (6.1 m)

$$R_I = \sqrt{\frac{\left(28,800 \, \frac{ft^3}{day}\right)(20.9\% - 5\%)}{\pi(20 \text{ ft})(4 \, \%/day)(0.25)}}$$

Therefore, the radius of influence at this site is approximately equal to 85 ft (26 m).

In practice, it is best to estimate the radius of influence from both pressure measurements and oxygen utilization. This incorporates all three of the key factors: pressure connection, air flow, and oxygen utilization. We have never encountered a site where this combined approach has overestimated the radius of influence.

The most conclusive determination of radius of influence is empirical measurement. The blower can be started and oxygen levels measured in monitoring points. The problem with this approach is that at a minimum, several days are required to reach steady state. At some sites, more than 30 days are required.

Well spacing typically is 1 to 1.5 times the radius of influence. When multiple wells are installed, some consideration may be given to airflow patterns. In theory, airflow lines may develop such that "dead zones" are created. However, given vertical and horizontal flow paths and diffusion, these dead zones are unlikely to occur, and we do not recommend routinely compensating for them.

IV. BLOWERS AND BLOWER SIZING

A blower provides the driving force to move air through the bioventing system. In selecting the blower size, one must consider the required air flowrate and the total system pressure drop. System pressure drop includes (1) the backpressure due to the vent wells and formation in an air injection configuration (or the vacuum induced in the wells and formation in an extraction configuration) plus (2) any pressure drop in the system piping and off-gas treatment system. This section describes the procedure for sizing a blower and uses a specific example for illustration purposes.

The two basic types of blowers are centrifugal machines and positive displacement machines. Positive displacement blowers are further subdivided into rotating machines and reciprocating machines (Figure 6-14). Selection of the appropriate type and size is based on the airflow requirement and the suction and discharge pressures presented to the

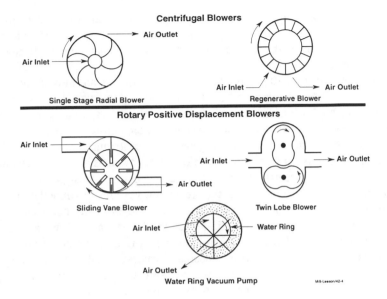

Figure 6-14. Schematic Diagram of Blower Types

blower during operation at the design air flowrate. Centrifugal blowers generally are favored when air flow requirements are high and/or the system pressure drop is low. Rotating positive displacement blowers generally provide lower airflow capacity and higher pressures than centrifugal blowers, but can generate moderate-to-high vacuum at the blower inlet. Due to their vacuum capability, rotating positive displacement blowers may be used for systems operating in an extraction configuration. Reciprocating positive displacement machines typically are used for applications requiring very high pressure. Except for single action diaphragm pumps used for soil gas sampling, reciprocating positive displacement pumps rarely are used in bioventing applications and are not discussed further. The required pressure or vacuum in the well is a function of the soil gas permeability, which is determined through field tests as described in Chapter 5.

A. CENTRIFUGAL BLOWERS

Centrifugal blowers impart kinetic energy to the air stream by means of a rapidly rotating impeller or propeller. Part of the added kinetic energy then is converted to pressure head in the blower casing as the fluid leaves the impeller. Examples of centrifugal blowers include radial blowers, regenerative radial blowers, multistage radial blowers, and axial blowers.

In a radial blower, air enters at the center of the housing and is picked up by an impeller vane near the axis of rotation (low-velocity area). Air is pushed radially away from the axis of rotation and accelerated by the impeller vane. Air exits the tip of the vane at high speed and enters the volute casing where the air velocity drops, converting kinetic energy into pressure head.

Regenerative centrifugal blowers provide efficient air movement in the flowrate and pressure drop ranges encountered in soil vapor extraction and bioventing applications and can produce moderate vacuum at the suction port. They are available in nonsparking, explosion-proof designs. As a result of these capabilities, the regenerative centrifugal blower is widely used in soil vapor extraction and bioventing systems. Unlike standard, single-stage radial centrifugal blower, the regenerative design uses a short-bladed turbine impeller. As the regenerative blower impeller rotates, centrifugal acceleration moves the air from the base of the blade to the blade tip. As the fast-moving air leaves the blade tip, it flows around the housing contour and back down to the base of the next blade where the flow pattern is repeated. This repeated acceleration allows a regenerative blower to produce higher differential pressure than a conventional, single-stage radial flow design. The regenerative blowers can also produce higher vacuum at the suction port in comparison with a pure radial flow design but are not able to reach the high-vacuum conditions provided by rotary positive displacement blowers.

B. ROTARY POSITIVE DISPLACEMENT BLOWERS

Rotary positive displacement blowers impart energy to the air stream by means of a rotating element displacing a fixed volume with each revolution. Examples of rotary positive displacement blowers include twin lobe blowers, water ring vacuum pumps, sliding vane blowers, and flexible vane blowers. Sliding vane and flexible vane blowers may be used for soil gas sampling or other low-flow applications but have too low an airflow capacity to act as the air handler in a bioventing system. Lobe blowers and water ring vacuum pumps have both seen some service in soil vapor extraction and bioventing systems where moderate-to-high vacuum is needed.

In a twin-lobe blower, two figure-eight-shaped lobe impellers are mounted on parallel shafts and rotate in opposite directions. As each impeller lobe passes the pump inlet, it traps a volume of gas and carries it around the case to the pump outlet. The rotation speed of the two impellers is controlled so that the volume created at the inlet side of the casing is larger than the volume at the outlet side of the casing, resulting in compression of the air trapped by the impeller lobe.

A water ring vacuum pump uses a rotating vaned impeller in a cylindrical pump casing. The impeller axis of rotation is off center with respect to the pump housing. A uniformly thick layer of water is formed on the inside of the pump casing by the rotary action of the impeller.

Since the impeller is off-center, the cavity formed between two impeller vanes and the water seal changes size as the vanes move around the pump housing. Air enters the pump where the cavity formed by the vanes and the water seal is large and is discharged where the cavity is small, thus increasing the pressure of the pumped gas.

C. BLOWER SELECTION AND SIZING

Proper sizing and selection of a blower is essential to ensure that the unit can deliver the required airflow at the necessary pressure and that it operates properly. Choosing the wrong blower can result in an inability to deliver sufficient oxygen or a significantly shortened blower life. It is best to select the blower to allow operation near the middle of its performance range. A blower operating near its maximum pressure/vacuum is running inefficiently and under stressed conditions, thereby increasing operating costs and shortening its life. Selection of an oversized blower reduces operating efficiency and increases capital costs unnecessarily. Example 6-4 illustrates a typical decision process for selection and sizing of a blower.

Example 6-4. Selection and Sizing of a Blower: For the site described in Example 6-2, we will need to deliver 16 cfm (453 L/min) of air to the example treatment cell. Based on the soil gas permeability test conducted at the site, operating pressures of $10"H_2O$ (2.5×10^3 Pa) were required to deliver 16 cfm (453 L/min). A regenerative air blower is selected as the blower of choice because it operates efficiently at the specified flowrate and pressure. Blower performance curves were obtained for three different size blowers (0.1, 0.125, and 2.5 hp, respectively), all of which might be expected to produce 16 cfm (453 L/min). The curves are shown in Figure 6-15.

The performance curves indicate that Blower #1 is too small and would not be able to provide 16 cfm (453 L/min) at $10"H_2O$ (2.5×10^3 Pa). Although blower #3 could provide 16 cfm (453 L/min) at $10"H_2O$ (2.5×10^3 Pa), it would be operating at the lower end of performance and would be too big. The performance curve for blower #2 shows that it would be a good choice. Blower #2 is rated to deliver as much as 21 cfm (595 L/min) at $10"H_2O$ (2.5×10^3 Pa). The excess air flow can be bypassed to the atmosphere, allowing adjustment for the 16 cfm (453 L/min) flow into the vent. If volatilization is not a concern and the additional air flow is not a problem, the entire flow can be injected into the vent well.

The example described above is a simplified case that shows how to select and size a blower for use in bioventing. Situations in the field may become more complicated if there are significant seasonal variations in soil gas permeability or other parameters affecting gas flow and oxygen demand. The key design consideration is to select and size a blower for the most demanding conditions, i.e., when oxygen demand is highest and soil gas permeability is lowest. Incorporating a bypass into the system plumbing will allow for reducing airflow delivered to the soil. The operating principles of several blower types are outlined in the following

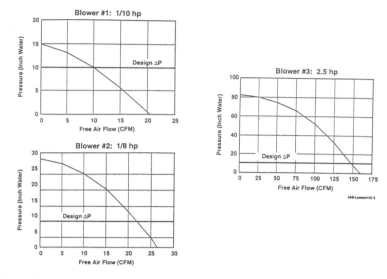

Figure 6-15. Performance Curves for Three Different Size Blowers (1/10, 1/8, and 2.5 hp, respectively)

sections. Further information on pumps and blowers may be found in *Pumping Manual* (1989) and *Pump Handbook* (Karassik et al., 1991)[1].

V. VENT WELL CONSTRUCTION

Vent well construction is fairly standard, and general guidelines are provided here. If existing groundwater monitoring wells at the site are screened above the water table, these can be used as vent wells. This option is appropriate for air-injection systems but will be less successful for air extraction systems because the applied vacuum will cause a rise in the water table that could submerge the screened interval.

The diameter of the vent well typically is between 2 and 4 inches (5.1 to 10 cm), although larger and smaller diameters have been used successfully. Vent well diameter will depend on the soil type, ease of drilling, and the area and depth of the contaminated volume. In most shallow or sandy soils, a 2-inch-diameter (5.1-cm) vent well will provide adequate airflow for bioventing. For sites with contamination extending below 30 ft (9.1 m) or in low permeability soils, a 3- or 4-inch (7.6- or 10-cm) vent well is recommended since this will allow for greater airflow to aerate a greater

---

[1] Refer to Appendix B for recommended specifications and manufacturers for the blowers.

volume.   As the well depths increase, the fractional cost of well-construction materials per ft of well decrease significantly[1].

The vent well typically is constructed of schedule 40 polyvinyl chloride (PVC), and should be screened with a slot size that maximizes airflow through the soil.   The screened interval should extend through as much of the contaminated profile as possible, with the bottom of the screen corresponding to the lowest historical level of the water table.   When designing the screen for an extraction well, the potential for water table upconing must be taken into account.   If the bottom of the screened interval is close to the water table, water will be pulled into the vent well, reducing its effectiveness.   If it is necessary to screen below the water table, additional screened length above the water table may be necessary to offset water table upconing.

Hollow-stem auguring is the most common drilling method; however, a solid-stem auger is acceptable in more cohesive soils.   The AFCEE is also investigating the use of cone penetrometer (CPT) wells for bioventing. Many other drilling techniques also are appropriate.   In shallow, softer soils, hand-auguring may be feasible.   Whenever possible, the diameter of the borehole should be at least two times greater than the vent well outside diameter.   The annular space corresponding to the screened interval should be filled with silica sand or equivalent.   The annular space above the screened interval should be sealed with a bentonite-and-grout slurry to prevent short-circuiting of air to or from the surface.   The construction detail of a typical vent well is shown in Figure 6-16.

To maintain the integrity of the vent well seal, as a rule of thumb, do not allow injection pressures measured in water depth to exceed the total grouted and sealed length.   For example, in a well with 3 ft (0.91 m) of bentonite seal and 3 ft (0.91 m) of grout, we would not exceed an injection pressure of $72''H_2O$ (1.8×10^4 Pa).   High pressures also can damage seals.   If the injection pressure exceeds the bearing capacity of the soil, fracturing is possible.   Care must be taken with injection wells to ensure that a good seal has been obtained.   Injection wells should be installed with a bentonite-and-grout slurry.   Dry bentonite chips do not provide an adequate seal unless the chips are hydrated continuously during installation.

VI. MONITORING POINT CONSTRUCTION

Soil gas monitoring points are used for pressure and soil gas measurements and are a very important component of a bioventing system. Proper construction of monitoring points is essential for monitoring

---

[1] Refer to Appendix B for recommended specifications and manufacturers for the vent well construction materials.

122

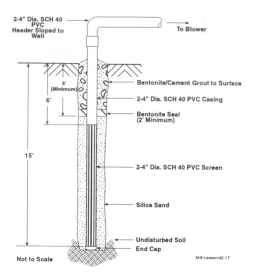

Figure 6-16. Schematic Diagram of a Typical Vent Well Construction

localized pressure and soil gas concentrations. To the extent possible, the monitoring points must be located in contaminated soils with greater than 1,000 mg/kg of total petroleum hydrocarbon. If monitoring points are not located in contaminated soil, meaningful in situ respiration data cannot be collected.

In addition, monitoring points should be located with consideration given to soil gas permeability testing and radius of influence determination. Monitoring points should be located at varying distances from the vent well. The distances from the vent well will vary depending on soil type; suggested monitoring point spacing is shown in Table 6-3.

In practice, each monitoring point cluster usually is screened to at least three depths. The deepest screen should be placed either at or near the bottom of contamination if a water table is not encountered, or a minimum of 2 to 3 ft (0.61 to 0.91 m) above the water table if it is encountered. Consideration should be given to potential seasonal water table fluctuations and soil type in finalizing the depth. In more permeable soil, the monitoring point can be screened closer to the water table. In less permeable soil, it must be screened further above the water table. The shallowest screen normally will be 3 to 5 ft (0.91 to 1.5 m) below land surface. The intermediate screen should be placed at a reasonable interval at a depth corresponding to the center-to-upper one-fourth of the vent well screen. In some cases, it may be desirable to add additional screened depths to more fully monitor the contaminated interval, to monitor differing

Table 6-3
Recommended Spacing for Monitoring Points

Soil Type	Depth to Top of Vent Well Screen (ft)[1]	Spacing Interval (ft)[2]
Coarse Sand	5	5-10-20
	10	10-30-50
	>15	20-30-70
Medium Sand	5	10-20-30
	10	15-25-45
	>15	20-40-70
Fine Sand	5	10-20-40
	10	15-30-50
	>15	20-40-60
Silts	5	10-20-40
	10	15-30-50
	>15	20-40-60
Clays	5	10-20-30
	10	10-20-40
	>15	10-25-50

[1] This assumes 10 ft of vent well screen. If more screen is used, the >15-ft spacing should be used. [2] Monitoring point intervals are based on a venting flowrate range of 1 cfm per ft of screened interval for clays and up to 3 cfm per ft of screened interval for coarse sands.

stratigraphic intervals, or to adequately monitor deeper sites with broadly screened vent wells.

Example 6-5. Selection of Depth Intervals for Monitoring Points: Site soils are sandy with groundwater at 30 ft (9.1 m). The vent well was screened from 17.5 to 27.5 ft (5.3 to 8.4 m) below land surface. Therefore, monitoring point depth intervals chosen were 28 ft (8.5 m), 22.5 ft (6.9 m), and 3 ft (0.91 m). For sites with vent wells deeper than 30 ft (9.1 m), more depths may be screened, depending on stratigraphy.

Monitoring point construction will vary depending on the drilling depth and technique. The monitoring points consist of a small-diameter (¼-inch [0.64 cm]) tube to the specified depth, with a screen approximately 6 inches (15 cm) long and 1 inch (2.5 cm) in diameter. In shallow open-hole installations, rigid tubing (i.e., schedule 80 ¼-inch [0.64-cm]) PVC terminating in the center of a gravel or sand pack may be adequate. The gravel or sand pack normally should extend for an interval of 1 to 2 ft (0.30 to 0.61 m), with the screen centered. In low-permeability soils, a larger gravel pack may be desirable. In wet soils, a longer gravel pack with the screen near the top may be desirable. A bentonite seal at least 2 ft (0.31 m) thick normally is required above and below the gravel pack. Figure 6-17 shows the construction detail of a typical monitoring point installation.

For relatively shallow installations in more permeable soils, a hand-driven system may be used. In such a system, a sacrificial drive point with Tygon™, Teflon™, or other appropriate tubing is driven to the desired depth. Then the steel outer tubing is retrieved, leaving the drive point and the inner flexible tubing in place. Because this type of installation allows little or no sand pack or seal placement, it should be used only in relatively permeable soils where sample collection will not be a problem or in soils that will "self heal" to prevent short-circuiting. Surface completion of the hand-driven points should be the same as for those installed in borings.

Monitoring points typically are used to collect soil gas for carbon dioxide and oxygen analysis in the 0 to 25% range, and for hydrocarbons greater than 100 ppmv. The tubing material must have sufficient strength and be nonreactive, appropriate materials include nylon and Tygon™. Sorption and gas interaction with the tubing materials have not been significant problems for this application. If a monitoring point will be used to monitor specific organics in the low-ppm or ppb range, Teflon™ or stainless steel may be necessary. However, this normally will not be the case.

A sufficient number of monitoring points should be installed to ensure representative sampling. The actual number installed is site-specific and is driven primarily by plume size, the cost of installing and monitoring additional monitoring points, and the scope of the project. If air injection is being considered in the bioventing test, a nest of monitoring points must be located between the vent well and any buildings that may be at risk to ensure that they are well beyond the radius of influence or that vapor phase hydrocarbons are biodegraded before air reaches the structure. Consider installing approximately one monitoring point for every four vent wells installed, with a minimum of three monitoring points for a given site.

Temperature monitoring typically is conducted by attaching thermocouples to monitoring points. Type J or K thermocouples can be used and should be attached to the monitoring point depth of interest. In general, soil temperatures vary little across a site, but do vary with depth to the ground surface. Therefore, few thermocouples are required for adequate soil temperature monitoring at a given site.

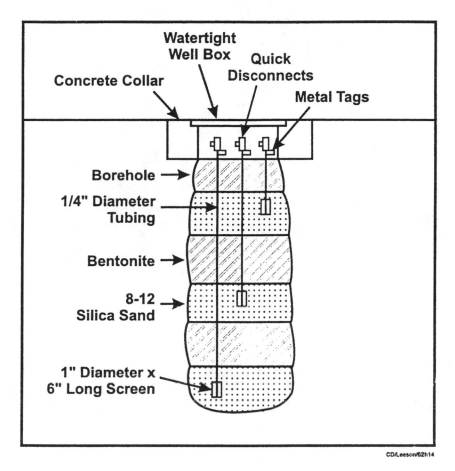

Figure 6-17. Schematic Diagram of a Typical Monitoring Point Construction

CHAPTER 7

BIOVENTING IMPLEMENTATION:
PERFORMANCE MONITORING

The following sections provide suggestions for monitoring bioventing systems. These methods provide a means of tracking the performance of a bioventing system over time. Methods discussed include soil gas sampling, in situ respiration testing, biodegradation and volatilization quantification, surface emissions measurement, optional monitoring, and operation and maintenance of the bioventing system. These methods are discussed in the following sections.

I. SOIL GAS MONITORING

Periodic soil gas monitoring should be conducted to ensure that the bioventing site is well-oxygenated[1]. Initially, soil gas should be monitored weekly until the site becomes fully aerated. Once full aeration is achieved, the bioventing system operation can be optimized. After this initial period, soil gas monitoring normally is conducted semiannually for the first year during the warmest and coldest months and annually thereafter. If it is not possible to conduct an in situ respiration test during different seasons, then it should be conducted under similar conditions as the initial test. Due to the relative simplicity of most bioventing systems, frequent soil gas monitoring rarely is necessary to ensure proper operation.

II. IN SITU RESPIRATION TESTING

In situ respiration testing should be conducted periodically as a means of monitoring the progress of site remediation. As site remediation progresses and contaminant concentrations are reduced, in situ respiration rates should approach those measured in the uncontaminated area. It is not necessary to conduct frequent in situ respiration tests. In situ respiration tests normally are conducted quarterly for the first year and annually thereafter.

In situ respiration tests for performance are conducted somewhat differently than the test for site characterization described in Chapter 5. During system operation, an in situ respiration test is conducted by first measuring soil gas concentrations of oxygen, carbon dioxide, and total

---

[1] Refer to Chapter 5 for more detail on sampling and analyzing soil gases.

hydrocarbons during system operation. After these measurements are collected, the bioventing system is turned off and soil gas monitoring is conducted periodically to measure oxygen disappearance and carbon dioxide production. No inert tracer gas is added at this time because the initial testing should have determined whether diffusion or monitoring point leakage was occurring. Calculation of biodegradation rates is accomplished in the same manner as described in Chapter 5.

In situ respiration testing should be used as the primary indicator for site closure. A good indication that the site is remediated and that final soil sampling can be conducted is when the in situ respiration rate in the contaminated area is similar to that in the uncontaminated area. In situ respiration testing to determine remediation success is preferable economically to relying on soil sampling as the sole indicator of site remediation, because it eliminates the high cost of intermediate soil sampling.

In situ respiration rates can be expected to vary with time. Generally, temperature is the most significant driver of short-term (within one year) changes. Over longer periods, contaminant reduction will reduce rates. One phenomenon frequently observed is a substantial decline in rates from the initial in situ respiration rates to subsequent measurements. It appears that this generally is due to placement of monitoring points in less-contaminated soils. NAPL contamination usually is distributed in a very heterogeneous manner. Under nonventing conditions, volatilization will spread hydrocarbons in soil gas, resulting in more heterogeneous contamination. However, the soil contaminated in this fashion has a much lower total concentration, because the sorbed hydrocarbons are present at much lower levels than in soils that actually contain NAPLs. If a monitoring point is placed in soil having only sorbed and vapor-phase contamination, the initial rates will be high. However, remediation will rapidly reduce the sorbed concentrations and the in situ respiration rates will fall quickly, often by a factor of 5 to 10 in a few months. One indication of this is a low-rate apparent first-order oxygen decay curve, resulting in misleading rate data. It is difficult to eliminate this problem, but it can be limited by attempting to place monitoring points in the most highly contaminated soil.

III. QUANTIFICATION OF BIODEGRADATION AND VOLATILIZATION OF HYDROCARBONS DURING EXTRACTIVE BIOVENTING

Biodegradation and volatilization of hydrocarbons can be quantified during extractive bioventing through direct measurement of off-gas concentrations of oxygen and carbon dioxide. Bioventing systems that are operating in injection mode have been reconfigured briefly in order to provide these data. It should be noted, however, that in the case of injection mode systems, reconfiguration to extraction mode will provide an overestimate of the mass of hydrocarbons volatilized because the size of the in situ bioreactor is reduced[1].

The mass of hydrocarbons volatilized can be calculated as follows:

$$HC_{vol} = \frac{C_{V,HC}}{10^6} \times Q \times \rho_{hexane} \times MW_{hexane} \times \frac{kg}{1,000g} \times \frac{1,440 \ min}{day} \qquad (7\text{-}1)$$

where: HC_{vol} = mass of hydrocarbons volatilized (kg/day)

$C_{V,HC}$ = concentration of hydrocarbons in extracted off-gas (ppmv)

Q = flowrate (L/min or cfm)

ρ_{hexane} = density of hexane (moles/L)

MW_{hexane} = molecular weight of hexane (g/mole)

The mass of hydrocarbons biodegraded can be calculated as follows:

$$HC_{bio} = \left(\frac{C_{V,bkgd} - C_{V,O_2}}{100}\right) \times Q \times C \times \rho_{O_2} \times MW_{O_2} \times \frac{kg}{1,000 \ g} \times \frac{1,440 \ min}{day} \qquad (7\text{-}2)$$

where: HC_{bio} = mass of hydrocarbons biodegraded (kg/day)

$C_{V,bkgd}$ = concentration of oxygen in background, uncontaminated area (%)

$C_{V,O2}$ = concentration of oxygen in extracted off-gas (%)

C = mass ratio of hydrocarbon to oxygen degraded based on stoichiometry[2] (1/3.5)

Example 7-1 illustrates these calculations.

Example 7-1. Calculation of Volatilization and Biodegradation of Contaminants During Extraction: At a site undergoing extraction, concentrations of oxygen and TPH in the extracted soil gas at steady state are 19% and 140 ppmv, respectively. The system is operating at a flowrate of 4 cfm (113 L/min). Background oxygen concentrations are consistently at 20.9%. We first wish to calculate the mass of hydrocarbons volatilized.

Given the following parameters:

$C_{V,HC}$ = 140 ppmv
Q = 4 cfm (113 L/min)
ρ_{hexane} = 0.042 moles/L
MW_{hexane} = 84 g/mole

---

[1] Refer to Chapter 6 for a discussion of these issues.

[2] Refer to Chapter 5 for a discussion of stoichiometry.

130

Using Equation 7-1:

$$HC_{vol} = \left(\frac{140 \text{ ft}^3 \text{ hexane}}{10^6 \text{ ft}^3 \text{ air}}\right)\left(4\frac{\text{ft}^3}{\text{min}} \times \frac{28.3 \text{ L}}{\text{ft}^3}\right)\left(0.042\frac{\text{mole}}{\text{L}}\right)\left(84\frac{\text{g}}{\text{mole}}\right)\left(\frac{\text{kg}}{1,000 \text{ g}}\right)\left(\frac{1,440 \text{ min}}{\text{day}}\right)$$

Solving, the mass of hydrocarbons volatilized is 0.081 kg/day (0.18 lb/day).

To calculate the mass of hydrocarbons biodegraded, we use Equation 3-2:

$$HC_{bio} = \left(\frac{20.9 - 19.0}{100}\right)\left(4\frac{\text{ft}^3}{\text{min}} \times \frac{28.3 \text{ L}}{\text{ft}^3}\right)\left(\frac{1 \text{ g HC}}{3.5 \text{ g O}_2}\right)\left(0.042\frac{\text{mole}}{\text{L}}\right)\left(32\frac{\text{g}}{\text{mole}}\right)\left(\frac{\text{kg}}{1,000 \text{ g}}\right)\left(\frac{1,440 \text{ min}}{\text{day}}\right)$$

Solving, the mass of hydrocarbons biodegraded is approximately 1.2 kg/day (2.6 lb/day), or nearly an order of magnitude greater than the amount volatilized.

The fraction of total removal by biodegradation will be larger for injection systems because the opportunity for biodegradation is greater. In an injection mode, the vapors are pushed through the contaminated zone into the uncontaminated zone, allowing for additional biodegradation. However, when the system is operated in extraction mode, much of the vapor is removed from the soil before biodegradation can occur.

IV. SURFACE EMISSIONS SAMPLING

Surface emissions sampling is not necessary at most bioventing sites. Under the U.S. Air Force Bioventing Initiative, it was conducted at only 5 of 125 sites to quantify volatilization of contaminants attributed to air injection. Although surface emissions typically do not occur or are very low at bioventing sites due to low air flowrates, possible surface emissions often are a regulatory concern and surface emission rates may need to be quantified in order to obtain regulatory approval for bioventing. However, it should be noted that, according to the U.S. EPA document *Estimation of Air Impacts for Bioventing Systems Used at Superfund Sites* (U.S. EPA, 1993, EPA 451/R-93-003), emissions from bioventing sites operating in an injection mode are thought to be minimal. Therefore, they are not discussed in this document.

One standard surface emission sampling protocol using isolation flux chamber procedures is described in Dupont and Reineman (1986) and Dupont (1988) and is illustrated in Figure 7-1. The system consists of a square Teflon™ box that covers a surface area of approximately 0.45 m². The box is fitted with inlet and outlet ports for the entry and exit of high-purity air. Inside the box is a manifold that delivers the air supply uniformly across the soil surface. The same type of manifold is fitted to the exit port of the box. This configuration delivers an even flow of air across the entire soil surface under the box to generate a representative sample.

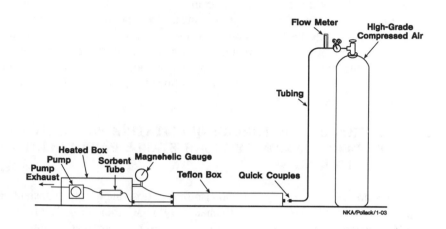

Figure 7-1. Schematic Diagram of a Surface Emissions Monitoring Device

The air exiting the Teflon™ box is directed to a sampling box that contains a sorbent tube and a pump. Also attached to the box is a purge line that accommodates the excess flow from the Teflon™ box that is not drawn into the sorbent tube. A Magnehelic™ gauge is used to indicate if a zero pressure is being maintained on the entire system.

In all cases, a totally inert system is employed. Teflon™ tubing and stainless steel fittings assure that there is no contribution to or removal of organics from the air stream. The pump is located on the back side of the sorbent trap so that it is not in a position to contaminate the sample flow.

To calculate the actual emission rates of organic compounds from the soil surface into the atmosphere, the following formula for dynamic enclosure techniques is employed (McVeety, 1991):

$$F = \frac{C_v \, V_r}{A} \qquad (7\text{-}3)$$

where:
F = flux in mass/area-time (g/m²-min)

C_v = the concentration of the gas in units of mass/volume (g/m³)

V_r = volumetric flowrate of sweep gas (m³/min)

A = soil surface area covered by enclosure (m²)

At bioventing sites where surface emissions have been measured, BTEX and TPH surface emission rates have been several orders of magnitude below regulatory levels.   As an example, Table 7-1 illustrates surface emissions results from six bioventing sites.   In general, surface emissions are very low, with TPH emission rates less than 1 lb/day.   These emission rates are well below most regulatory limits and illustrate that properly designed bioventing systems create no significant air emissions.   These results provide strong support for continued operation of bioventing systems in injection mode.

V.   OPTIONAL MONITORING: QUALITATIVE VALIDATION OF BIODEGRADATION THROUGH STABLE CARBON ISOTOPE MONITORING

Measurement of stable carbon isotope ratios may help substantiate biodegradation (Aggarwal and Hinchee, 1991).   Carbon dioxide produced by hydrocarbon degradation may be distinguished from that produced by other processes based on the carbon isotopic compositions characteristic of the source material and/or the fractionation accompanying microbial metabolism (Suchomel et al., 1990; Stahl, 1980; McMahon et al., 1990). As shown in Figure 7-2, carbon dioxide generated from natural organic material has a $\delta^{13}C$ of approximately -10 to -15, whereas carbon dioxide generated from petroleum hydrocarbons has a $\delta^{13}C$ of approximately -20 to -30.   This measurement is not required to validate biodegradation, since the in situ respiration test is used for this purpose; therefore, it should be conducted only if dictated by regulatory concerns.

VI. OPERATION AND MAINTENANCE

Bioventing systems are very simple, with minimal mechanical and electrical parts.   If the system is operated in an injection mode, a simple visual system check to ensure that the blower is operating within its intended flowrate, pressure, and temperature range is required.   Weekly system checks are desirable.   These system checks often can be conducted by someone on site because little technical knowledge of the process is required.   Minor maintenance such as replacing filters, flow meters, or gauges may be necessary.

If an extraction system or an extraction/reinjection bioventing system is installed, more intensive maintenance is likely to be required.   Extraction systems have knockout drums that require draining and treatment of condensate.   In addition, in the case of extraction-only systems, off-gas may need to be monitored regularly to ensure that emissions are within regulatory guidelines.   Any off-gas treatment system also will require periodic checks to ensure proper operation.

Table 7-1
Surface Emissions Sampling at Bioventing Sites

Base	Site Type	Air Injection Depth (ft)	Air Injection Rate (cfm)	Area of Influence (ft^2)	Total Flux Estimate (lb/day)
Beale AFB, CA	Fire Training Pit	10 - 25	30	6,500	0.15
Bolling AFB, D.C.	Diesel Spill	10 - 15	20	5,100	0.44
Eielson AFB, AK	JP-4 Spill	6.5 - 13	30.0	43,600	0.011
Fairchild AFB, WA	JP-4 Spill	5 - 10	15	5,100	0.33
McClellan AFB, CA	JP-4 Spill	10 - 55	50	9,700	0.066
Plattsburgh AFB, NY	Fire Training Pit	10 - 35	13	11,500	0.44

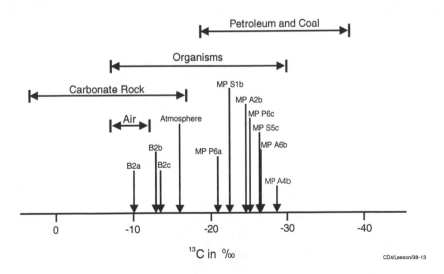

Figure 7-2. Carbon Isotopic Compositions of Soil Gas Carbon Dioxide at Site 20, Eielson AFB, Alaska, August 1993

134

Blowers used for bioventing systems typically last for several years and should not need replacement. To date, two bioventing systems have been operating for 3 years with the original blower in place (Battelle, 1994; Leeson et al., 1995). Of the 125 blowers installed to date under the U.S. Air Force Bioventing Initiative, only three have required repair or replacement.

CHAPTER 8

BIOVENTING IMPLEMENTATION:
PROCESS EVALUATION/SITE CLOSURE

I. IN SITU RESPIRATION TESTING

In situ respiration testing should be used as the primary indicator for site closure. As discussed in Chapter 5, as site remediation progresses and contaminants are degraded, the measured in situ respiration rates will approach background respiration rates. When the in situ respiration rate in the contaminated area approaches that in the uncontaminated area, this is a good indication that the site is remediated and final soil sampling can be conducted. Initially, one can estimate the time necessary for cleanup of the site based on in situ respiration rates as shown in Example 8-1.

Example 8-1. Calculation of Remediation Time Based on In Situ Respiration Rates:
For this example, we assume an average oxygen utilization rate of 6% O_2/day and an initial average soil concentration of 6,000 mg TPH/kg soil. Oxygen utilization is related to hydrocarbon degradation by the following equations:

$$C_6H_{14} + 9.5O_2 \rightarrow 6CO_2 + 7H_2O \qquad (8\text{-}1)$$

$$k_B = -0.68\ k_o \qquad (8\text{-}2)$$

Using the above assumptions, an oxygen utilization rate of 6% O_2/day would correspond to a biodegradation rate of approximately 4.1 mg/kg-day. Given that the initial soil concentration is 6,000 mg/kg, an estimate of cleanup time is calculated as follows:

$$\frac{C_o}{k_B} = \text{Cleanup time} \qquad (8\text{-}3)$$

$$\frac{6,000\ \text{mg/kg}}{4.1\ \text{mg/kg-day}} = 1,500\ \text{days} \approx 4\ \text{years}$$

This calculation provides a reasonable "ball park" estimate of the amount of time necessary to remediate the site. This method tends to underestimate treatment time because k_B decreases over time. At the same time, this calculation overestimates treatment time because it does not consider treatment in the expanded bioreactor. Therefore, the calculation

135

must be coupled with process monitoring to provide field-based evidence that the site actually is remediated within this time period. Due to widely variable contaminant concentrations, the average biodegradation rate does not reflect actual biodegradation rates throughout the site. Biodegradation rates also may fluctuate with season and as contaminant concentrations decrease. Therefore, process monitoring is an important parameter in determining treatment time.

II. SOIL SAMPLING

Soil sampling should not be used as a process-monitoring technique. Due to the inherently high variability of hydrocarbons in soils, the number of samples required to produce a meaningful result is prohibitive until contamination levels approach 90 to 99% cleanup. The amount of soil sampling conducted at a site has a tremendous impact on the cost of the project. Minimizing soil sampling will make a remediation effort much more cost effective. With bioventing systems, in situ respiration testing can indicate when the site is clean and therefore when to collect final soil samples. The number of final soil samples collected usually is driven by regulatory issues. The Department of Natural Resources of the State of Michigan published a guidance document for verification of soil remediation. This document provides several methods for statistical sampling strategies (Department of Natural Resources, MI, 1994). This document provides information on design of the sampling grid and determination of the upper confidence limit (UCL) of the final mean. The upper confidence limit is calculated from the following equation:

$$UCL = \overline{X} + [t = 0.95(n - 1)] \, S_X \qquad (8\text{-}4)$$

where:
UCL	=	upper confidence limit
X	=	average contaminant concentration
bracketed term	=	one-tailed t-test at n-1 degrees of freedom (see Table 8-1 for values)
S_X	=	standard error of the mean, which is calculated as follows:

$$S_X = \frac{S}{\sqrt{n}} \qquad (8\text{-}5)$$

where: S = standard deviation
n = sample size

Table 8-1
Cumulative t Distribution

	0.550	0.750	0.080	p *0.900*	0.950	0.975	0.990	0.995
one- tailed	0.550	0.750	0.080	*0.900*	0.950	0.975	0.990	0.995
two- tailed	0.100	0.500	0.600	0.800	0.900	0.950	0.980	0.990
1	0.158	1.000	1.376	*3.078*	6.314	12.706	31.821	63.657
2	0.142	0.816	1.061	1.886	2.920	4.303	6.925	9.925
3	0.137	0.765	0.978	1.638	2.353	3.182	4.541	5.841
4	0.134	0.741	0.941	1.533	2.132	2.776	3.747	4.604
5	0.132	0.727	0.920	1.476	2.015	2.571	3.365	4.032
6	0.131	0.718	0.906	1.440	1.943	2.447	3.143	3.707
7	0.130	0.711	0.896	1.415	1.895	2.365	2.998	3.499
8	0.130	0.706	0.889	1.397	1.860	2.306	2.896	3.355
9	0.129	0.703	0.883	1.383	1.833	2.262	2.821	3.250
10	0.129	0.700	0.879	1.372	1.812	2.228	2.764	3.169
11	0.129	0.697	0.876	1.363	1.796	2.201	2.718	3.106
12	0.128	0.695	0.873	1.356	1.782	2.179	2.681	3.055
13	0.128	0.694	0.870	1.350	1.771	2.160	2.650	3.012
14	0.128	0.692	0.868	1.345	1.761	2.145	2.624	2.977
15	0.128	0.691	0.866	1.341	1.753	2.131	2.602	2.947

138

If the calculated upper confidence limit is higher than the regulatory threshold, than the lambda relationship is used to calculate the appropriate sample size:

$$\lambda = \frac{RT - \overline{X}}{S} \qquad (8\text{-}6)$$

where: λ =   statistical parameter (see Table 8-2 for values)
       RT   =   regulatory threshold
       X   =   average contaminant concentration
       S   =   standard deviation

Once λ is calculated by referring to Table 8-2, the number of additional samples required to verify cleanup can be determined, as is shown in Example 8-2.

Example 8-2. Statistical Evaluation of Contaminant Data: At this site, three preliminary soil samples were collected to estimate a sample mean and standard deviation. The initial sample mean was 90 mg/kg TPH with a standard deviation of 30 mg/kg. The regulatory threshold is 100 mg/kg TPH. Calculating the UCL:

$$UCL = 90 + (2.920) \times \left(\frac{30}{\sqrt{3}}\right) = 141 \ mg/kg$$

Given that this value is above the regulatory threshold, the lambda calculation is performed to determine how many additional samples are required to verify cleanup.

$$\lambda = \frac{100 - 90}{30} = 0.33$$

From Table 8-2, for $\alpha = 0.05$ and $\beta = 0.05$, a sample size of between 90 and 122 additional samples is required.

An alternative method for estimating final sample size is provided by Ott (1984). This method determines the number of soil samples required to show a statistical difference between initial and final contaminant concentrations.

$$n = \frac{\sigma^2 \left(z_\alpha + z_\beta\right)^2}{\left(\mu_o - \mu\right)^2} \qquad (8\text{-}7)$$

Table 8-2
Number of Observations for t Test of Mean

Single-sided Double-sided		Level for t test									
		α = 0.01 α = 0.02					α = 0.05 α = 0.10				
λ	β	0.01	0.05	0.1	0.2	0.5	0.01	0.05	0.1	0.2	0.5
0.05											
0.10											
0.15											122
0.20						139					70
0.25						90			139	101	45
0.30					115	63		122	97	71	32
0.35				109	85	47		90	72	52	24
0.40			101	85	66	37	101	70	55	40	19
0.45		110	81	68	53	30	80	55	44	33	15
0.50		90	66	55	43	25	65	45	36	27	13
0.55		75	55	46	36	21	54	38	30	22	11
0.60		63	47	39	31	18	46	32	26	19	9
0.65		55	41	34	27	16	39	28	22	17	8
0.70		47	35	30	24	14	34	24	19	15	8
0.75		42	31	27	21	13	30	21	17	13	7
0.80		37	28	24	19	12	27	19	15	12	6
0.85		33	25	21	17	11	24	17	14	11	6
0.90		29	23	19	16	10	21	15	13	10	5
0.95		27	21	18	14	9	19	14	11	9	5
1.00		25	19	16	13	9	18	13	11	8	5

where: n   =   number of final soil samples to collect
        σ^2 =   population variance of the initial soil sampling event
        z_α =   probability of a Type I error
        z_β =   probability of a Type II error
        μ_o =   mean of the initial soil sampling event
        μ =   estimated mean of the final soil sampling event

As the difference between the initial and final means increases, the number of samples required to show a statistical difference between the two sampling events decreases. As shown in Table 8-3, as hydrocarbons are further degraded, fewer soil samples are required to show a statistical difference in the two means. This concept is illustrated in Example 8-3.

Example 8-3. Calculation of Final Number of Soil Samples for Site Closure: At this site, 83 initial soil samples were collected with a mean TPH concentration of 6,000 mg/kg and a standard deviation of 8,000 mg/kg (typical of many bioventing sites). The average biodegradation rate at this site was 4.1 mg/kg-day. Given that the system has been operating for 3.5 years, we can estimate the final mean TPH concentration as follows:

$$4.1 \text{ mg/kg-day} \times 1{,}278 \text{ days} = 5{,}240 \text{ mg/kg TPH degraded}$$

$$\text{Estimated final[TPH]} = 6{,}000 \text{ mg/kg} - 5{,}240 \text{ mg/kg} = 760 \text{ mg/kg}$$

Using this estimate of the final mean TPH concentration, the number of samples needed to provide statistically significant data can be calculated. Using Equation 8-7 and the following parameters:

σ = $(8{,}000)^2$
z_α =   1.645 (for $\alpha = 0.05$)
z_β =   2.33 (for $\beta = 0.01$)
μ_o =   6,000 mg/kg
μ =   525 mg/kg

Selected z values are shown in Table 8-4. The z_α and z_β are found by finding areas corresponding to $(0.5-\alpha)$ and $(0.5-\beta)$, respectively.

$$n = \frac{(8{,}000)^2(1.645 + 2.33)^2}{(6{,}000 - 760)^2}$$

Therefore, the number of final soil samples that must be collected is 37.

Table 8-3
Calculation of the Number of Samples Required to Show a Statistical
Difference Between Means of Two Sampling Events

Time From Initiation of Bioventing (days)	Estimated Amount of Hydrocarbon Degraded (mg/kg)	Estimation Amount of Hydrocarbon Remaining (mg/kg)	Number of Samples Required
180	1440	4560	731
365	2920	3080	178
540	4320	1680	81
730	5840	160	44

Table 8-4
Selected z Values for Estimation of Final Soil Sample Number (Ott, 1984)[1]

z	.00	.01	.02	.03	.04	.05
0.0	.0000	.0040	.0080	.0120	.0160	.0199
0.1	.0398	.0438	.0478	.0517	.0557	.0596
0.2	.0793	.0832	.0871	.0910	.0948	.0987
0.3	.1179	.1217	.1255	.1293	.1331	.1368
0.4	.1554	.1591	.1628	.1664	.1700	.1736
0.5	.1915	.1950	.1985	.2019	.2054	.2088
0.6	.2257	.2291	.2324	.2357	.2398	.2422
0.7	.2580	.2611	.2642	.2673	.2704	.2734
0.8	.2881	.2910	.2939	.2967	.2995	.3023
0.9	.3159	.3186	.3212	.3238	.3264	.3289
1.0	.3413	.3438	.3461	.3485	.3508	.3531
1.1	.3643	.3665	.3686	.3708	.3729	.3749
1.2	.3849	.3869	.3888	.3907	.3925	.3944
1.3	.4032	.4049	.4066	.4082	.4099	.4115
1.4	.4192	.4207	.4222	.4236	.4251	.4265
1.5	.4332	.4345	.4357	.4370	.4382	.4394
1.6	.4452	.4463	.4474	.4484	.4495	.4505
1.7	.4554	.4564	.4573	.4582	.4591	.4599
1.8	.4641	.4649	.4656	.4664	.4671	.4678
1.9	.4731	.4719	.4726	.4732	.4738	.4744
2.0	.4772	.4778	.4783	.4788	.4793	.4798
2.1	.4821	.4826	.4830	.4734	.4838	.4842
2.2	.4861	.4864	.4868	.4871	.4875	.4878
2.3	.4893	.4896	.4898	.4901	.4904	.4906

z	.00	.01	.02	.03	.04	.05
2.4	.4918	.4920	.4922	.4925	.4927	.4829
2.5	.4938	.4940	.4941	.4943	.4945	.4846

[1] Bolded areas correspond to determining z_β. Italicized areas correspond to determining z_α.

CHAPTER 9

COSTS

Based on Air Force and recent commercial applications of this technology, the total cost of in situ soil remediation using the bioventing technology is $10 to $60 per cubic yard (Downey et al., 1994b). At sites with over 10,000 cubic yards of contaminated soil, costs of less than $10 per cubic yard have been achieved. Costs greater than $60 per cubic yard are associated with smaller sites, but bioventing still can offer significant advantages over more disruptive excavation options. Operation and maintenance costs are minimal, particularly when on-site personnel perform the simple system checks and routine maintenance that are needed. Table 9-1 provides a detailed cost breakdown of remediation of 5,000 cubic yards of soil contaminated with an average concentration of 3,000 mg of JP-4 jet fuel per kg of soil.

Ward (1992) compared costs of bioventing to other in situ bioremediation technologies (Table 9-2). Costs shown in Table 9-2 reflect actual costs for these three technologies at fuel spills at Traverse City, Michigan. Even though the area treated through bioventing was larger than that treated with hydrogen peroxide or nitrate, total costs for bioventing were significantly lower than for the other technologies.

Figure 9-1 provides a comparison of estimated unit costs for several technologies commonly used for remediation of fuel-contaminated soils. All costs are based on the treatment of soil contaminated with 3,000 mg JP-4 jet fuel per kg of soil. Costs are provided for the following remediation scenarios: two years of in situ bioventing; excavation and one year of on-base landfarming with leachate controls; one year of soil vapor extraction with thermal vapor treatment; and excavation followed by low-temperature thermal desorption. The cost of reconstructing excavated areas is not included. At many sites with contamination beneath concrete and buildings, bioventing is the only cost-effective treatment option available.

146

Table 9-1
Typical Full-Scale Bioventing Costs (Downey et al., 1994b)

Task	Total Cost ($)
Site Visit/Planning	5,000
Work Plan Preparation	6,000
Pilot Testing	27,000
Regulatory Approval	3,000
Full-Scale Construction	
Design	7,500
Drilling/Sampling[1]	15,000
Installation/Startup	4,000
Two-Year Monitoring	6,500
Two-Year Power	2,800
Soil Sampling at 2 Years	13,500
Total	**90,300**

[1] Assumes four air injection wells drilled to a depth of 15 ft.

Table 9-2
Cost Comparison of In Situ Bioremediation Technologies Utilized at Fuel Spill Sites
(Ward et al., 1992)

| Task | Total Costs ($ per m^3 of Contaminated Earth) | | |
	Hydrogen Peroxide	Nitrate	Bioventing[1]
Construction[2]	45	118	26
Labor/Monitoring	72	96	40
Chemicals	500	30	0.44
Electricity	24	12	6.8
Total	**641**	**256**	**73**

[1] Values reflect only first 4 months of demonstration. [2] Prorated to a 5-year service life on buildings, pumps, and blowers.

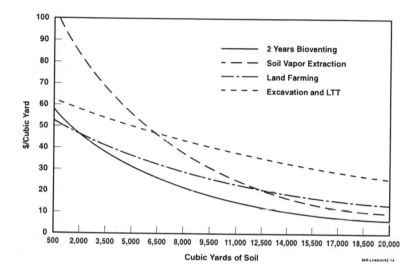

Figure 9-1. Comparison of Costs for Various Remedial Technologies for Fuel-Contaminated Soils (Downey et al., 1994b)

CHAPTER 10

REFERENCES

Aggarwal, P.K. and R.E. Hinchee. 1991. "Monitoring In Situ Biodegradation of Hydrocarbons by Using Stable Carbon Isotopes." *Env. Science and Tech.*, *25*:1178-1180.

Aggarwal, P.K., J.L. Means, and R.E. Hinchee. 1991. "Formulation of Nutrient Solutions for In Situ Bioremediation." In: R.E. Hinchee and R.F. Olfenbuttel (Eds.), *In Situ Bioreclamation*, Butterworth-Heinemann, Stoneham, MA. pp. 51-66.

Alleman, B.C., R.E. Hinchee, R.C. Brenner, and P.T. McCauley. 1995. "Bioventing PAH Contamination at the Reilly Tar Site." In: R.E. Hinchee, R.N. Miller, and P.C. Johnson (Eds.), *In Situ Aeration: Air Sparging, Bioventing, and Related Remediation Processes*, Battelle Press, Columbus, OH. pp. 473-482.

American Petroleum Institute. 1987. *Field Study of Enhanced Subsurface Biodegradation of Hydrocarbons Using Hydrogen Peroxide as an Oxygen Source*. API Publication #4448, Washington, D.C., 1987.

Anonymous. 1986. "In Situ Reclamation of Petroleum Contaminated Sub-Soil by Subsurface Venting and Enhanced Biodegradation." *Research Disclosure*, No. 26233, 92-93.

Atlas, R.M. 1986. "Microbial Degradation of Petroleum Hydrocarbons: An Environmental Perspective." *Microbiol. Rev.*, *45*:180-209.

Baker, J.N., D.A. Nickerson, and P.R. Guest. 1993. "Use of a Horizontal Air-Dispersion System to Enhance Biodegradation of Diesel Fuel Contaminated Sites." In: *Proceedings of the 1993 Petroleum Hydrocarbons and Organic Chemicals in Ground Water: Prevention, Detection, and Restoration*. Presented by the American Petroleum Institute and the Association of Ground Water Scientists and Engineers. Water Well Journal Publishing Company, Dublin, OH.

Battelle. 1994. "Bioremediation of Hazardous Wastes at CERCLA and RCRA Sites: Hill AFB 280 Site, Low-Intensity Bioreclamation." Report

150

Prepared by Battelle Memorial Institute for the U.S. Environmental Protection Agency, January, 1994.

Bennedsen, M.B., J.P. Scott, and J.D. Hartley. 1987. "Use of Vapor Extraction Systems for In Situ Removal of Volatile Organic Compounds from Soil." In: *Proceedings of National Conference on Hazardous Wastes and Hazardous Materials*, Washington, DC. pp. 92-95.

Bilbo, C.M., E. Arvin, and H. Holst. 1992. "Modelling the Growth of Methane-Oxidizing Bacteria in a Fixed Biofilm." *Water Research*, 26:301-309.

Braker, W. and A.L. Mossmon. 1980. *Matheson Gas Data Book,* 6th ed., Lyndhurst, NJ.

Britton, L.N. 1985. *Field Studies on the Use of Hydrogen Peroxide to Enhance Microbial Degradation of Gasoline.* API Publication #4389, Washington, D.C., 1985.

Brock, T.D., D.W. Smith, and M.T. Madigan. 1984. *Biology of Microorganisms,* 4th ed., Prentice-Hall, Inc., Englewood Cliffs, NJ.

Brown, R.A. and J.R. Crosbie. 1994. "Oxygen Sources for *In Situ* Bioremediation." *Bioremediation: Field Experience.* Lewis Publishers, Boca Raton, FL. pp. 311-331.

Brown, R.A., R.D. Norris, and R.L. Raymond. 1984. "Oxygen Transport in Contaminated Aquifers." In: *Proceedings of the NWWA/API Conference on Petroleum Hydrocarbons and Organic Chemicals in Groundwater*, National Water Well Association, Columbus, OH, November 1984.

Conner, A.T. 1989. "Case Study on Soil Venting." *Pollution Engineering*, 20(8):74-78.

Department of Natural Resources, MI. 1994. *Guidance Document, Verification of Soil Remediation*, Report Prepared by the Environmental Response Division and Waste Management Division of the Department of Natural Resources, MI. April, 1994.

DeVinny, J.S., L.G. Everett, J.C.S. Lu, and R.L. Stollar. 1990. *Subsurface Migration of Hazardous Wastes.* Van Nostrand Reinhold.

Downey, D.C. and J.F. Hall. 1994. Addendum One to Test Plan and Technical Protocol for a Field Treatability Test for Bioventing — Using Soil Gas Surveys to Determine Bioventing Feasibility and Natural Attenuation Potential. U.S. Air Force Center for Environmental Excellence, Brooks Air Force Base, TX.

Downey, D.C., R.E. Hinchee, M.S. Westray, and J.K. Slaughter. 1988. "Combined Biological and Physical Treatment of a Jet Fuel-Contaminated Aquifer." In: *Proceedings of the NWWA/API Conference on Petroleum Hydrocarbons and Organic Chemicals in Groundwater*, National Water Well Association, Columbus, OH, November 1988. pp. 627-645.

Downey, D.C., C.J. Pluhar, L.A. Dudus, P.G. Blystone, R.N. Miller, G.L. Lane, and S. Taffinder. 1994a. "Remediation of Gasoline-Contaminated Soils Using Regenerative Resin Vapor Treatment and In Situ Bioventing." In: *Proceedings of the 1994 Petroleum Hydrocarbons and Organic Chemicals in Ground Water: Prevention, Detection, and Restoration Conference.*

Downey, D.C., J.F. Hall, R.N. Miller, A. Leeson, and R.E. Hinchee. 1994b. Bioventing Performance and Cost Summary. U.S. Air Force Center for Environmental Excellence, Brooks Air Force Base, TX, February, 1994.

Downey, D.C., R.A. Frishmuth, S.A. Archabal, C.J. Pluhar, P.G. Blystone, and R.N. Miller. 1995. "Using In Situ Bioventing to Minimize Soil Vapor Extraction Costs." In: R.E. Hinchee, R.N. Miller, and P.C. Johnson (Eds.), *In Situ Aeration: Air Sparging, Bioventing, and Related Remediation Processes*, Battelle Press, Columbus, OII. pp. 247-266.

Dragun, J. 1988. "Microbial Degradation of Petroleum Products in Soil." In: E.J. Calabrese and P.T. Kostecki (Eds.), *Soils Contaminated by Petroleum: Environmental and Public Health Effects*. John Wiley & Sons, Inc., New York, NY. pp. 289-300.

Dupont, R.R. and J.A. Reineman. 1986. Evaluation of the Volatilization of Hazardous Constituents at Hazardous Waste Land Treatment Sites. U.S. EPA Office of Research and Development, R.S. Kerr Environmental Laboratory, Ada, OK. EPA 600/2-86/071.

Dupont, R.R. 1988. "A Sampling System for the Detection of Specific Hazardous Constituent Emissions From Soil Systems." In: *Hazardous Waste: Detection Control Treatment*. Amsterdam, The Netherlands, Elsevier Science Publisher, B.V. pp. 581-592

152

Dupont, R.R., W.J. Doucette, and R.E. Hinchee. 1991. "Assessment of *In Situ* Bioremediation Potential and the Application of Bioventing at a Fuel-Contaminated Site." In: R.E. Hinchee and R.F. Olfenbuttel (Eds.), *In Situ Bioreclamation: Applications and Investigations for Hydrocarbon and Contaminated Site Remediation.* Butterworth-Heinemann, Stoneham, MA. pp. 262-282.

Ely, D.L. and D.A. Heffner. 1988. "Process for In-Situ Biodegradation of Hydrocarbon Contaminated Soil." U.S. Patent Number 4,765,902.

Engineering-Science. 1992. Interim Pilot Test Results Report for Installation Restoration Program Site 3, Fire Training Area, Battle Creek ANGB, Michigan, Report prepared for the U.S. Air Force Center for Environmental Excellence, Brooks AFB, Texas.

Engineering-Science. 1994a. Interim Pilot Test Results Report for Site ST-35 Fuel Pumphouse No. J3 and Site ST-36 Underground Fuel Line Davis-Monthan AFB, Arizona. Report prepared for the U.S. Air Force Center for Environmental Excellence, Brooks AFB, Texas.

Engineering-Science. 1994b. Interim Pilot Test Results Report for Three Bioventing Sites at Patrick AFB and Cape Canaveral Air Force Station, Florida. Report prepared for the U.S. Air Force Center for Environmental Excellence, Brooks AFB, Texas.

Foor, D.C. and R.E. Hinchee. 1993. *Long-Term Bioventing System Installation at Tank #7 and Tank #5 of the Alyeska Valdez Marine Terminal, Valdez, Alaska*, Report prepared for America North/EMCON, Anchorage, AK, November 1993.

Hinchee, R.E. and M. Arthur. 1991. "Bench-Scale Studies of the Soil Aeration Process for Bioremediation of Petroleum Hydrocarbons." *J. Appl. Biochem. Biotech.*, *28/29*:901-906.

Hinchee, R.E. and S.K. Ong. 1992. "A Rapid In Situ Respiration Test for Measuring Aerobic Biodegradation Rates of Hydrocarbons in Soil." *Journal Air Waste Management Assoc.*, *42*(10):1305-1312.

Hinchee, R.E. and G. Smith. 1991. Unpublished data. Battelle, Columbus, OH.

Hinchee, R.E., D.C. Downey, and T. Beard. 1989a. "Enhancing Biodegradation of Petroleum Hydrocarbons Fuels in the Vadose Zone Through Soil Venting." In: *Proceedings of the 1993 Petroleum*

Hydrocarbons and Organic Chemicals in Ground Water: Prevention, Detection, and Restoration, presented by The American Petroleum Institute and The Association of Ground Water Scientists and Engineers. Water Well Journal Publishing Company, Dublin, OH. pp. 235-248.

Hinchee, R.E., D.C. Downey, J.K. Slaughter, D.A. Selby, M. Westray, and G.M. Long. 1989b. HQ AFESC/RDVW ESK-TR-88-78. Enhanced Bioreclamation of Jet Fuels: A Full-Scale Test at Eglin Air Force Base, Florida. ESL/TR/88-78. Air Force Engineering and Services Center.

Hinchee, R.E., S.K. Ong, R.N. Miller, D.C. Downey, and R. Frandt. 1992. Test Plan and Technical Protocol for a Field Treatability Test for Bioventing, Rev. 2. U.S. Air Force Center for Environmental Excellence, Brooks Air Force Base, TX.

Hinchee, R.E., D.C. Downey, and P. Aggarwal. 1991a. "Use of Hydrogen Peroxide as an Oxygen Source for In Situ Biodegradation: Part I. Field Studies." *J. Hazardous Materials,* 27:287-299.

Hinchee, R.E., D.C. Downey, R.R. Dupont, P. Aggarwal, and R.N. Miller. 1991b. "Enhancing Biodegradation of Petroleum Hydrocarbon through Soil Venting." *J. Hazardous Materials,* 27:315-325.

Hopkins, G.D., J. Munakata, L. Semprini, and P.L. McCarty. 1993. "Trichloroethylene Concentration Effects on Pilot Field-Scale In-Situ Groundwater Bioremediation by Phenol-Oxidizing Microorganisms." *Environ. Sci. Technol.,* 27(12):2542-2547.

Howard, P.H., R.S. Boethling, W.F. Jarvis, W.M. Meylan, and E.M. Michalenko. 1991. *Handbook of Environmental Degradation Rates.* Lewis Publishers, Chelsea, MI.

Huling, S.G., B.E. Bledsoe, and M.V. White. 1990. *Enhanced Biodegradation Utilizing Hydrogen Peroxide as a Supplemental Source of Oxygen: A Laboratory and Field Study.* EPA/600-290-006. 48 pp.

Johnson, P.C., M.W. Kemblowski, and J.D. Colthart. 1990. "Quantitative Analysis for the Cleanup of Hydrocarbon-Contaminated Soils by In-Situ Soil Venting." *Ground Water,* 28(3), May-June.

Johnson, P.C., C.C. Stanley, M.W. Kemblowski, D.L. Byers, and J.D. Colthart. 1990. "A Practical Approach to the Design, Operation, and Monitoring of In Situ Soil-Venting Systems." *Groundwater Monitoring Rev.,* 10(2):159-178.

154

Kampbell, D. 1993. "U.S. EPA Air Sparging Demonstration at Traverse City, Michigan." In: Environmental Restoration Symposium. Sponsored by the U.S. Air Force Center for Environmental Excellence, Brooks AFB, TX.

Karassik, I.J., W.C. Krutsch, W.H. Fraser, and J.P. Messina. 1991. *Pump Handbook,* pp. 2-202,3-70-99

Kittel, J.A., R.E. Hinchee, and M. Raj. 1995. Full-Scale Startup of a Soil Venting-Based In Situ Bioremediation Field Pilot Study at Fallon NAS, Nevada. Report prepared for the Naval Facilities Engineering Services, Port Hueneme, CA, February, 1994.

Leeson, A. and R.E. Hinchee. 1994. Field Treatability Study at the Greenwood Chemical Superfund Site, Albemarle County, Virginia. Report prepared for the U.S. Environmental Protection Agency, Cincinnati, OH, January, 1994.

Leeson, A., R.E. Hinchee, J.A. Kittel, E.A. Foote, G. Headington, and A. Pollack. 1995. Bioventing Feasibility Study at the Eielson AFB Site. Report prepared for the Environics Directorate of the Armstrong Laboratory, Tyndall AFB, FL, September, 1995.

Lide, D.R. and H.V. Kehianian. 1994. *CRC Handbook of Thermophysical and Thermochemical Data.* CRC Press, Boca Raton, FL.

Lund, N.-Ch., J. Swinianski, G. Gudehus, and D. Maier. 1991. "Laboratory and Field Tests for a Biological *In Situ* Remediation of a Coke Oven Plant." In: R.E. Hinchee and R.F. Olfenbuttel (Eds.), *In Situ Bioreclamation: Applications and Investigations for Hydrocarbon and Contaminated Site Remediation.* Butterworth-Heinemann Publishing Company, Stoneham, MA. pp. 396-412.

Lyman, W.J., P.J. Reidy, and B. Levy. 1992. *Mobility and Degradation of Organic Contaminants in Subsurface Environments.* C.K. Smoley, Inc., Chelsea, MI.

McMahon, P.B., D.F. Williams, and J.T. Morris. 1990. *Ground Water* 28:693-702.

McVeety, B.D. 1991. Current and Developing Analytical Technologies for Quantifying Biogenic Gas Emissions. Report on Contract 600/3-91/044 from Pacific Northwest Laboratory to U.S. Environmental Protection Agency, Corvallis, Oregon, NTIS Report #PB91-216523, June.

Metcalf and Eddy. 1979. *Wastewater Engineering Treatment, Disposal, Reuse.* McGraw-Hill, New York, NY. 920 pp.

Miller, R.N. 1990. "A Field Scale Investigation of Enhanced Petroleum Hydrocarbon Biodegradation in the Vadose Zone Combining Soil Venting as an Oxygen Source with Moisture and Nutrient Additions." Ph.D. Dissertation. Utah State University, Logan, UT.

Miller, R.N. and R.E. Hinchee. 1990. "A Field Scale Investigation of Enhanced Petroleum Hydrocarbon Biodegradation in the Vadose Zone - Tyndall AFB, Florida." In: *Proceedings NWWA/API Conference on Petroleum Hydrocarbons and Organic Chemicals in Ground Water.* Houston, TX. 1990.

Miller, R.N., C.C. Vogel, and R.E. Hinchee. 1991. "A Field-Scale Investigation of Petroleum Hydrocarbon Biodegradation in the Vadose Zone Enhanced by Soil Venting at Tyndall AFB, Florida." In: R.E. Hinchee and R.F. Olfenbuttel (Eds.), *In Situ Bioreclamation.* Butterworth-Heinemann, Stoneham, MA. pp. 283-302.

Miller, R.N., D.C. Downey, V.A. Carmen, R.E. Hinchee, and A. Leeson. 1993. "A Summary of Bioventing Performance at Multiple Air Force Sites." In: *Proceedings of the 1993 Petroleum Hydrocarbons and Organic Chemicals in Ground Water: Prevention, Detection, and Restoration.* November 10-12, Houston, TX.

Morgan, P. and R.J. Watkinson. 1992. "Factors Limiting the Supply and Efficiency of Nutrient and Oxygen Supplements for the In Situ Biotreatment of Contaminated Soil and Groundwater." *Water Res.,* 26(1):73-78.

Mormile, M.R., S. Liu, and J.M. Suflita. 1994. "Anaerobic Biodegradation of Gasoline Oxygenates: Extrapolation of Information to Multiple Sites and Redox Conditions." *Environmental Sci. Technol.,* 28(9):1727-1732.

Newman, B., M. Martinson, G. Smith, and L. McCain. 1993. "Enhanced Biodegradation of Hydrocarbons In-Situ via Bioventing." In: *Proceedings of the 1993 Petroleum Hydrocarbons and Organic Chemicals in Ground Water: Prevention, Detection, and Restoration.* November 10-12, Houston, TX. Presented by the American Petroleum Institute and the Association of Ground Water Scientists and Engineers. Water Well Journal Publishing Company, Dublin, OH.

Oak Ridge National Laboratory. 1989. Soil Characteristics: Data Summary, Hill Air Force Base Building 914 Fuel Spill Soil Venting Project. An unpublished report to the U.S. Air Force.

Ong, S.K., R. Hinchee, R. Hoeppel, and R. Scholze. 1991. "In Situ Respirometry for Determining Aerobic Degradation Rates." In: R.E. Hinchee and R.F. Olfenbuttel (Eds.), *In Situ Bioreclamation Applications and Investigations for Hydrocarbon and Contaminated Site Remediation.* Boston, MA. pp. 541-545.

Ong, S.K., A. Leeson, R.E. Hinchee, J. Kittel, C.M. Vogel, G.D. Sayles, and R.N. Miller. 1994. "Cold Climate Application of Bioventing," In: R.E. Hinchee, B.C. Alleman, R.E. Hoeppel, and R.N. Miller (Eds.), *Hydrocarbon Bioremediation.* Lewis Publishers, Boca Raton, FL. pp. 444-453.

Ostendorf, D.W. and D.H. Kampbell. 1989. "Vertical Profiles and Near Surface Traps for Field Measurement of Volatile Pollution in the Subsurface Environment." In: *Proceedings of NWWA Conference on New Techniques for Quantifying the Physical and Chemical Properties of Heterogeneous Aquifers*, Dallas, TX. Westarp Wiss., Essen, Germany, pp. 475-485.

Ott, L. 1984. *An Introduction to Statistical Methods and Data Analysis.* 2nd ed. Duxbury Press, Boston, MA.

Peck, R.B., W.E. Hanson, and T.H. Thornburn. 1962. *Foundation Engineering,* Second Edition. John Wiley & Sons, New York, NY.

Phelps, M.B., F.T. Stanin, and D.C. Downey. 1995. In: R.E. Hinchee, R. N. Miller, and P.C. Johnson (Eds.), *In Situ Aeration: Air Sparging, Bioventing, and Related Remediation Processes*, Battelle Press, Columbus, OH. pp. 277-282.

Pumping Manual. 1989. Trade & Technical Press Ltd. Morden, Surrey, England. pp. 120, 122.

Reisinger, J., IST, Inc., Atlanta, GA. 1994. Personal communication.

Rivett, M.O. and J.A. Cherry. 1991. "The Effectiveness of Soil Gas Surveys in Delineation of Groundwater Contamination: Controlled Experiments at the Borden Field Site." In: *Proceedings of the 1991 Petroleum Hydrocarbons and Organic Chemicals in Ground Water: Prevention, Detection, and Restoration.* November. Houston, TX.

Sayles, G.D., R.C. Brenner, R.E. Hinchee, A. Leeson, C.M. Vogel, and R.N. Miller. 1994a. "Bioventing of Jet Fuel Spills I: Bioventing in a Cold Climate with Soil Warming at Eielson AFB, Alaska." In: *Symposium on Bioremediation of Hazardous Wastes: Research, Development and Field Applications*. EPA/600/R-94/075, June 1994, pp. 15-21.

Sayles, G.D., R.C. Brenner, R.E. Hinchee, and E. Elliott. 1994b. "Bioventing of Jet Fuel Spills II: Bioventing in a Deep Vadose Zone at Hill AFB, Utah." In: *Symposium on Bioremediation of Hazardous Wastes: Research, Development and Field Applications*. EPA/600/R-94/075, June 1994, pp. 22-28.

Schumb, W.C., C.N. Satterfield and R.L. WEntworth. 1955. *Hydrogen Peroxide*. Van Nostrand Reinhold, New York, NY.

Sellers, K. and C.Y. Fan. 1991. "Soil Vapor Extraction: Air Permeability Testing and Estimation Methods." In: *Proceedings of the 17th RREL Hazardous Waste Research Symposium*. EPA/600/9-91/002. April 1991.

Spain, J.C., J.D. Milligan, D.C. Downey and J.K. Slaughter. 1989. "Excessive Bacterial Decomposition of H_2O_2 During Enhanced Biodegradation." *J. Groundwater, 27*:163-167.

Staatsuitgeverij. 1986. Proceedings of a Workshop, 20-21 March, 1986. *Bodembeschermingsreeeks* No. 9; *Biotechnologische Bodemsanering*, pp. 31-33. Rapportnr. 851105002, ISBN 90-12-054133, Ordernr. 250-154-59; Staatsuitgeverij Den Haag: The Netherlands.

Stahl, W.J. 1980. *Geochim. Cosmochim. Acta, 44*:1903-1907.

Suchomel, K.H., D.K. Kreamer, and A. Long. 1990. *Environ. Sci. Technol., 24*:1824-1831.

Texas Research Institute. 1980. *Laboratory Scale Gasoline Spill and Venting Experiment*. American Petroleum Institute, Interim Report No. 7743-5:JST.

Texas Research Institute. 1984. *Forced Venting to Remove Gasoline Vapor from a Large-Scale Model Aquifer*. American Petroleum Institute, Final Report No. 82101-F:TAV.

158

Urlings, L.G.C.M., H.B.R.J. van Vree, and W. van der Galien. 1990. "Application of Biotechnology in Soil Remediation." *Envirotech Vienna.* pp. 238-251.

U.S. Environmental Protection Agency. 1993. "Engineering Issue — In Situ Bioremediation of Contaminated Unsaturated Subsurface Soils." Draft. EPA/540/S-93/501. Office of Research and Development, Robert S. Kerr Environmental Research Laboratory, Ada, OK; and Office of Solid Waste and Emergency Response, Washington, DC.

van Eyk, J. and C. Vreeken. 1988. "Venting-Mediated Removal of Petrol from Subsurface Soil Strata as a Result of Stimulated Evaporation and Enhanced Biodegradation." *Med. Fac. Landbouww. Riiksuniv. Gent,* 53(4b):1873-1884.

van Eyk, J. and C. Vreeken. 1989a. "Model of Petroleum Mineralization Response to Soil Aeration to Aid in Site-Specific, In Situ Biological Remediation." In: Jousma et al. (Eds.), *Groundwater Contamination: Use of Models in Decision-Making, Proceedings of an International Conference on Groundwater Contamination.* Kluwer, Boston/London. pp. 365-371.

van Eyk, J. and C. Vreeken. 1989b. "Venting-Mediated Removal of Diesel Oil from Subsurface Soil Strata as a Result of Stimulated Evaporation and Enhanced Biodegradation." In: *Hazardous Waste and Contaminated Sites, Envirotech Vienna,* Vol. 2, Session 3. ISBN 389432-009-5. Westarp Wiss., Essen, Germany. pp. 475-485.

van Eyk, J. 1994. "Venting and Bioventing for the In Situ Removal of Petroleum From Soil", In: R.E. Hinchee, B.C. Alleman, R.E. Hoeppel, and R.N. Miller (Eds.), *Hydrocarbon Bioremediation.* Lewis Publishers, Boca Raton, FL, pp. 243-251.

Wackett, L.P. and D.T. Gibson. 1992. "Degradation of Trichloroethylene by Toluene Dioxygenase in Whole-Cell Studies with *Pseudomonas putida* F1." *Appl. Environmental Microbiol., 54*(7):1703-1708.

Ward, C.H. 1988. "A Quantitative Demonstration of the Raymond Process for In-Situ Biorestoration of Contaminated Aquifers." In: *Proceedings of NWWA/API Conference on Petroleum Hydrocarbons and Organic Chemicals in Groundwater.* pp. 723-746.

Ward, C.H. 1992. "Performance and Cost Evaluation of Bioremediation Techniques for Fuel Spills." In: *Proceedings of In Situ Bioremediation Symposium '92.* pp. 15-21.

Wiedemeier, T.H., J.T. Wilson, D.H. Kampbell, R.N. Miller, and J.E. Hansen. 1995. Technical Protocol for Implementing Intrinsic Remediation with Long-Term Monitoring for Natural Attenuation of Fuel Contamination Dissolved in Ground Water. Report prepared for the U.S. Air Force Center for Environmental Excellence, Technology Transfer Division, Brooks Air Force Base, San Antonio, TX, April 1995.

Wilson, J.T. and C.H. Ward. 1986. "Opportunities for Bioremediation of Aquifers Contaminated with Petroleum Hydrocarbons." *J. Ind. Microbiol.*, 27:109-116.

Zachary, S.P. and L.G. Everett. 1993. "In-Situ Active/Passive Bioreclamation of Vadose Zone Soils Contaminated with Gasoline and Waste Oil Using Soil Vapor Extraction/Bioventing: Laboratory Pilot Study to Full Scale Site Operation." In: *Proceedings of the 1993 Petroleum Hydrocarbons and Organic Chemicals in Ground Water: Prevention, Detection, and Restoration* presented by The American Petroleum Institute and The Association of Ground Water Scientists and Engineers. Water Well Journal Publishing Company, Dublin, OH.

APPENDIX A

STATISTICAL ANALYSES OF
U.S. AIR FORCE BIOVENTING INITIATIVE DATA

In May 1992, the U.S. Air Force began the Bioventing Initiative to examine bioventing at 55 contaminated sites throughout the country. In December 1992, the program was increased to more than 130 sites due to increased demand by Air Force managers. To date, data have been collected from 125 contaminated sites at a total of 50 Air Force bases, one Army base, one Naval installation, and one Department of Transportation installation. Sites are located in 35 states and in all 10 U.S. EPA regions. The selected sites represent a wide range of contaminant types and concentrations, soil types, contaminant depths, climatic conditions, and regulatory frameworks. Sites were selected based on contamination level (preferably > 1,000 mg/kg TPH). The selections were not biased with regard to factors such as soil type or climatic conditions, in order to properly evaluate bioventing potential under both favorable and unfavorable conditions.

A Bioventing Test Protocol was developed which provided strict guidelines for treatability testing and bioventing system design. The Bioventing Test Protocol was peer reviewed and was reviewed by U.S. EPA Headquarters and the U.S. EPA National Risk Management Research Laboratory. Using the Bioventing Test Protocol, initial testing was conducted at each site to determine whether bioventing was feasible. Based on the initial testing, a decision was made whether to install a bioventing system for 1 year of operation. At the majority of sites (95%), a bioventing system was installed for the 1-year operational period. At the end of this time period, each Air Force base could either elect to keep the bioventing system in operation or remove it if the site was deemed to have been remediated sufficiently.

At each site in which a bioventing system was installed, a series of data was collected: initial site characterization data consisting of soil and soil gas sampling, in situ respiration rate testing results, and soil gas permeability testing results; 6-month in situ respiration testing results; and 1-year soil and soil gas sampling and in situ respiration testing results. Data from the initial testing have been used in the statistical analyses as described in Section II. A summary of the results to date with potential implications is presented in the following sections.

161

I. ESTIMATE OF CONTAMINANT REMOVAL AT U.S. AIR FORCE BIOVENTING INITIATIVE SITES

At all U.S. Air Force Bioventing Initiative sites in which a blower was installed and operated for 1 year, initial and final soil and soil gas BTEX and TPH concentrations were measured. The approach was to compile a limited number of samples from each site and statistically analyze for trends to avoid known spatial variability. Distributions of soil and soil gas BTEX and TPH concentrations from the initial and 1-year sampling events are shown in Figures A-1 through A-4, respectively. The average soil and soil gas BTEX and TPH concentrations across all sites are shown in Figure A-5. In general, the most dramatic reductions were observed in BTEX removal in both soil and soil gas samples. As an example, soil results from Site 3 at Battle Creek ANGB are shown in Figure A-6. After 1-year of bioventing operation the BTEX concentrations are very low and are no longer a source of groundwater contamination; therefore, site closure is now a viable option for this site.

The objective of the 1-year sampling event was not to collect the large number of samples required for statistical significance for a single site. Rather, the objective was to give a qualitative indication of changes in contaminant mass. Soil gas samples are somewhat similar to composite samples in that they are collected over a wide area. Thus, they provide an indication of changes in soil gas profiles (Downey and Hall, 1994). Blower operation was discontinued 30 days prior to sample collection to allow for soil gas equilibration. In contrast, soil samples are discrete point samples subject to large variabilities over small distances/soil types. Given this variability, coupled with known sampling and analytical variabilities, a large number of samples at a single site would have to be collected to conclusively determine real changes in soil contamination. Due to the limited number of samples, these results should not be viewed as conclusive indicators of bioventing progress or evidence of the success or failure of this technology.

If a risk-based approach to remediation is used which focuses on removing the soluble, mobile, and more toxic BTEX component of the fuel, remediation times can be significantly reduced. As discussed in the Tyndall AFB case history in Chapter 4, the BTEX fraction was removed preferentially over TPH. The potential for bioventing to preferentially remove BTEX makes this technology suitable for risk-based remediations. In addition, the low levels of BTEX that have been encountered at the majority of the U.S. Air Force Bioventing Initiative sites further support an emphasis on risk-based remediation (Figure A-7). Over 85% of the initial soil samples contained less than 1 mg/kg of benzene.

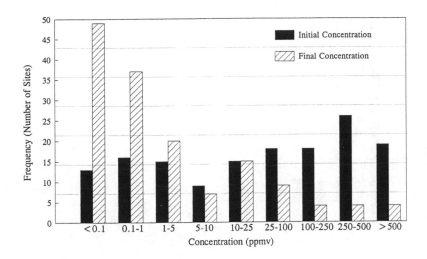

Figure A-1.   **Soil Gas BTEX Concentrations at Bioventing Initiative Sites: Initial and 1-Year Data**

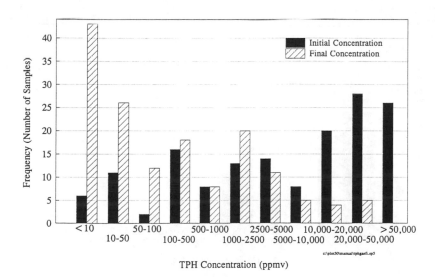

Figure A-2.   **Soil Gas TPH Concentrations at Bioventing Initiative Sites: Initial and 1-Year Data**

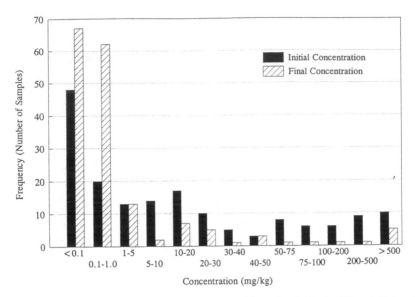

Figure A-3.   **Soil BTEX Concentrations at Bioventing Initiative Sites: Initial and 1-Year Data**

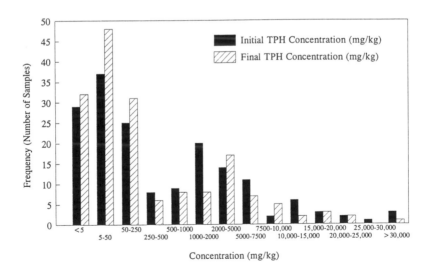

Figure A-4.   **Soil TPH Concentrations at Bioventing Initiative Sites: Initial and 1-Year Data**

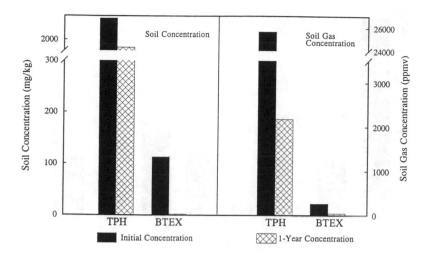

Figure A-5.   Average Soil and Soil Gas BTEX and TPH Concentrations at Bioventing Initiative Sites: Initial and 1-Year Data

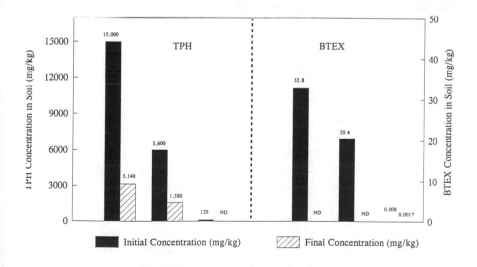

Figure A-6.   Initial and Final Soil Sampling Results at Site 3, Battle Creek ANGB, Michigan

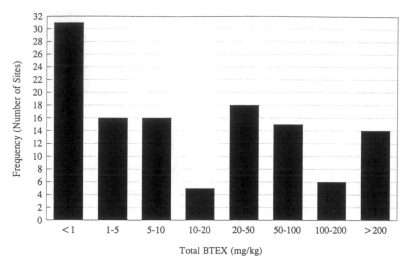

Figure A-7. Average BTEX Concentrations at Bioventing Initiative Sites

II. STATISTICAL ANALYSIS OF U.S. AIR FORCE BIOVENTING INITIATIVE DATA

One of the primary objectives of the U.S. Air Force Bioventing Initiative was to develop a large database of bioventing systems from which it could be determined which parameters are most important in evaluating whether to implement bioventing. This effort is the largest field effort to date where data have been collected in a consistent manner to allow for direct comparison of results across sites. Results of the statistical analyses can be used to evaluate which soil measurements should be taken and, if bioventing performance is poor, which parameters may be adjusted to improve performance.

Data generated from the U.S. Air Force Bioventing Initiative were subjected to thorough statistical analyses to determine which parameters most influenced observed oxygen utilization rates. Procedures used for conducting the statistical analyses and the results of these analyses are presented in the following sections.

A. PROCEDURES FOR STATISTICAL ANALYSIS

Data collected from 125 U.S. Air Force Bioventing Initiative sites have been analyzed for this study. The study involved in situ respiration test data, soil gas permeability test data, and soil chemistry and nutrient data from each site. Several parameters were measured in the soil samples. The statistical analyses had five specific objectives:

- To develop a consistent statistical approach for calculating the oxygen utilization and carbon dioxide production rates from the in situ respiration data.
- To characterize the oxygen utilization rate as a function of parameters measured during initial testing.
- To determine the relationship between carbon dioxide production rate and pH or alkalinity by characterizing the ratio of oxygen utilization rate to carbon dioxide production rate as a function primarily of pH and alkalinity.
- To characterize soil gas permeability as a function of particle size and moisture content.
- To compare TKN concentrations at contaminated sites with those at uncontaminated background areas.

Averages for oxygen utilization and carbon dioxide production rates and soil parameters were computed for each site. All subsequent analyses were performed on the site averages. Table A-1 displays the parameters included in the statistical analyses, their units, and transformations performed on these parameters whenever necessary.

Data were stored in Statistical Analysis System (SAS) databases, and all statistical manipulations and analyses were conducted using the SAS software package. Methods used for characterizing the data and the final regression model are presented in the following sections for each of the listed objectives.

B. CALCULATION OF OXYGEN UTILIZATION AND CARBON DIOXIDE PRODUCTION RATES

A statistical analysis was conducted to consistently calculate oxygen utilization and carbon dioxide production rates. A linear, time-related change in oxygen and carbon dioxide levels that is characterized by a constant (or zero-order) rate is typical of most of the sites. However, in some sites, a two-piecewise linear change is observed. An initial rapid rate is observed followed by a leveling off. This change in rates generally occurs once oxygen becomes limiting, typically below 5 to 10% oxygen.

The two-piecewise regression model, with a slope change at time t_0, was fitted to the oxygen (and carbon dioxide) versus time data obtained at every monitoring point. The piecewise regression model is presented below:

$$R_i = \alpha + \beta t_i \qquad t_i \leq t_0 \qquad\qquad \text{(A-1)}$$

168

Table A-1
Data Parameters Included in the Statistical Analyses

Category	Parameter	Units	Transformation[a]	Acronym[b]
In Situ Respiration Rates	Oxygen utilization rate	%/hr	Log	O2
	Carbon dioxide production rate	%/hr	None	CO2
	Ratio of the carbon dioxide production rate to oxygen utilization rate	No units	Square root	Ratio
Soil Parameters	Soil gas TPH	ppmv	Log	tphsg
	Soil gas BTEX	ppmv	Log	btexsg
	Soil TPH	mg/kg	Log	tphs
	Soil BTEX	mg/kg	Log	btexs
	pH	No units	Log	PH
	Alkalinity	mg/kg as CaCO$_3$	Log	ALK
	Iron content	mg/kg	Log	IRN
	Nitrogen content	mg/kg	Log	NIT
	Phosphorus content	mg/kg	Log	PHO
	Moisture content	% wt	None	MOI
	Gravel	% wt	None	GRA
	Sand	% wt	None	SAN
	Silt	% wt	None	SIL
	Clay[c]	% wt	None and log	CLA
	Soil gas permeability	Darcy	Log	PRM
	Soil temperature	Celsius	None	TMP
Other	Season (time of year)	Day	None	season

a = Transformation was applied to the parameter for purposes of statistical analysis. b = Acronym is used for the parameter in this report. c = The correlations in Figures A-9 through A-14 and Figure A-17 are based on untransformed clay.

$$R_i = (\alpha + \beta t_0) + (\beta + \delta)(t_i - t_0) \qquad t_i > t_0 \qquad \text{(A-2)}$$

for $i = 1, 2, \ldots,$ # of observations at each monitoring point, and where:

R_i =   measured i^{th} oxygen or carbon dioxide level at time t_i (%);

α =   oxygen or carbon dioxide level at initial time (%);

β =   rate of change of oxygen or carbon dioxide level with time (%/hr);

δ =   increase or decrease in the rate of change at time t_0 (%/hr);

t_0 =   time at which the slope change occurs (hr).

The piecewise regression model was implemented using the NLIN procedure (nonlinear regression procedure) in the SAS software package.

The parameter δ in the above model measures the increase or decrease in the slope at time t_0. Therefore, the statistical significance of δ confirmed the suitability of a two-piecewise model fitted to the data. The rate of oxygen utilization (or carbon dioxide production) was estimated from the slope of the first linear piece, β, whenever δ was statistically significant at the 0.05 significance level. For example, Figure A-8 presents the piecewise linear model fitted to oxygen data at a monitoring point at Site FSA-1, Air Force Plant (AFP) 4, Texas, where β was estimated to be -1.1%/hr.

In cases where δ was not significant at the 0.05 level, a linear regression model of the following form was fitted to the data:

$$R_i = \alpha + \beta t_i \qquad\qquad \text{for all } t_i \qquad \text{(A-3)}$$

where the rate of oxygen utilization (or carbon dioxide production) was determined from the slope of the straight line, β.

For cases in which six or fewer observations were available at a monitoring point, or when the oxygen levels exhibited virtually no change over a short initial time period followed by a linear change, the piecewise analysis was not attempted. In such cases, a linear regression model, as described above, was fitted. In these cases, the suitability of the linear model was confirmed by inspection of the model-fit to observed data.

170

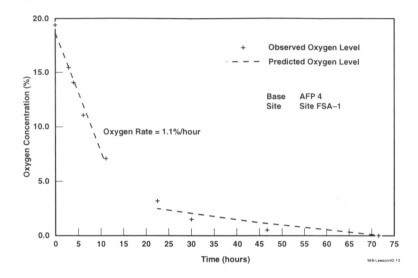

Figure A-8. Use of Piecewise Analysis of Oxygen Utilization Data from Site FSA-1, AFP 4, Texas

C. CORRELATION OF OXYGEN UTILIZATION RATES AND ENVIRONMENTAL PARAMETERS

A preliminary analysis of the untransformed data was performed in which a regression model was fitted to the oxygen utilization rate using forward stepwise regression. This model accounted for the effects of the soil parameters and their interactions. In order to reduce the effect of multicollinearity among the parameters on the fitted model, soil gas BTEX levels and gravel were excluded from the modeling. In other words, soil gas BTEX was highly correlated with soil gas TPH, and therefore it was concluded that the effect of soil gas BTEX levels on the oxygen utilization rate can almost completely be explained by soil gas TPH concentrations. Also, since the particle size levels added up to a constant value (100%), the effect of gravel was assumed to be redundant in the modeling.

As a result of fitting the regression model to the oxygen utilization rate it was found that soil particle sizes and permeability had a dominating influence on the oxygen utilization rate; that is, low levels of permeability and sand and high levels of silt and clay appeared to correlate strongly with high oxygen utilization rates.

In order to determine whether a handful of sites were unduly influencing the statistical modeling, sites with high oxygen utilization rates were examined in detail. Seven sites in the analyses had extremely high oxygen utilization rates, well above average rates from other sites. A two-sample t-test was performed on each parameter (e.g. sand, nitrogen, etc.) to determine whether the average value of the parameter over the seven sites was different from the corresponding average for the remaining sites. This analysis revealed statistically significant differences in particle size, soil gas permeability, and soil TPH concentrations between the two groups of sites (Table A-2). As a result of this analysis, it was determined that the seven sites with extremely high oxygen utilization rates were atypical with respect to their levels of particle size, soil gas permeability, and soil TPH concentrations.

In order to reduce the influence on the model for the oxygen utilization rate caused by these seven sites, the log transformation of the oxygen utilization rate was taken. Additionally, the log transform resulted in more normally distributed data for the oxygen utilization rate. However, sites with oxygen utilization rates near zero receive artificial importance as a result of the transformation. To eliminate this artificial effect caused by the log-transformation, all the log-transformed values of the oxygen utilization rate below −2.5 were censored, that is, set to a constant value of −2.5. Censoring was based on visual inspection of the log-transformed data.

Subsequently, the log transform of some of the soil parameters was taken if the data for the parameter were not well represented by a normal distribution. Normality in the data was checked using the Shapiro-Wilk test for normality and by observing histograms and normal probability plots.

As a preliminary step to determine the influence of the soil parameters on the oxygen utilization rate, correlations between the rate and each of the soil parameters were examined. This step was conducted to examine strong relationships between oxygen utilization rates and measured environmental parameters to assist in developing a statistical model describing performance at the U.S. Air Force Bioventing Initiative sites. First, the log transformation of the oxygen utilization rate and some of the soil parameters was taken to obtain more normally distributed data on each parameter (Table A-1). After these transformations, the data for each parameter were plotted against the corresponding data for each of the other parameters.

Figures A-9 through A-14 display the magnitude of the correlations among the data parameters. Specifically, Figures A-9 through A-11, display the correlations between the oxygen utilization rate and the soil parameters and Figures A-12 through A-14 present the correlations between the soil parameters. In each of these figures, ellipses are drawn on each plot containing 95% of the estimated bivariate distribution. The plots for which the ellipses are narrow represent pairs of elements that have a strong observed correlation. Pairs of elements that are positively correlated have the ellipse with the major axis running from the lower left

Table A-2
Parameters That Distinguish the Seven Sites
with High Oxygen Utilization Rates From the Remaining Sites

Parameter	Level of Parameter in Seven Sites Relative to Other Sites
Sand	Lower
Silt	Higher
Clay	Higher
Soil gas permeability	Lower
Soil TPH	Lower

to the upper right, whereas negative correlations are indicated by the major axis running from the lower right to the upper left. The magnitude of the correlation can be inferred from the shape of the ellipse by comparing it to the key figure. In the key figure, comparable ellipses are displayed for distributions with known correlations of 90%, 60%, 30%, and 0%.

For example, in Figures A-9 through A-11, it can be seen that the oxygen utilization rate is most positively correlated with nitrogen (Figure A-9, correlation coefficient $r=0.40$), moisture (Figure A-9, $r=0.30$), and soil gas TPH concentrations (Figure A-10, 0.20), and negatively correlated with sand (Figure A-11, $r=0.25$) and temperature (Figure A-10, $r=0.25$). These values indicate that high levels of nitrogen, moisture, and soil gas TPH concentrations and low levels of sand and temperature appear to correlate with high oxygen utilization rates.

It should be noted that the correlation between soil temperature and oxygen utilization rate is of little practical significance in this analysis. At a given site, temperature has been shown to correlate well with microbial activity, which displays peak activity in summer months and low activity in winter months. However, it has been noted that the temperature/microbial activity relationship is very site specific. In other words, microorganisms in Alaska will show peak activity in summer months with comparable oxygen utilization rates to organisms from more temperate climates; however, soil temperatures will be significantly different. Therefore, it is not possible to correlate rates with temperature under such different climatic conditions as were seen at U.S. Air Force Bioventing Initiative sites.

Among the soil parameters, the correlation coefficient between soil gas BTEX and TPH concentrations is 0.92 (Figure A-9) and that between pH and alkalinity is 0.75 (Figure A-9). The correlations between the particle sizes (sand, silt, and clay), moisture, and soil gas permeability are also pronounced.

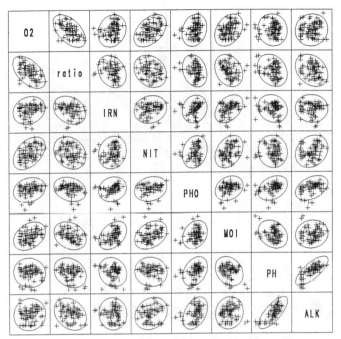

O_t = log O_e Rate   Ratio = $(CO_2$ Rate/O_2 Rate$)^{1/4}$ IRN = log Iron   NIT = log Nitrogen
PHO = log Phosphorus   MOI = Moisture   PH = log pH   ALK = log Alkalinity

Key to Correlation Scatterplots.

Figure A-9. Oxygen Utilization Rates, Oxygen:Carbon Dioxide Rate Ratios, Element Concentrations, Moisture Content, pH, and Alkalinity Site Average Correlation Scatterplot

174

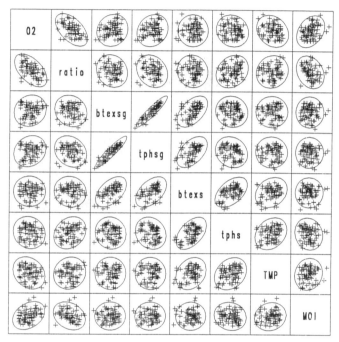

O_2 = log O_2 Rate   Ratio = $(CO_2 \text{ Rate}/O_2 \text{ Rate})^{\frac{1}{4}}$ btexsg = log BTEX in Soil Gas   tphsg = log TPH in Soil Gas
btexs = log BTEX in Soil   tphs = log TPH in Soil   TMP = Soil Temperature   MOI = Moisture

	90%	60%	30%	0%
Z1	⬭	⬭	⬭	◯

Key to Correlation Scatterplots.

Figure A-10.   Oxygen Utilization Rates, Oxygen:Carbon Dioxide Rate Ratios,
Contaminant Concentrations, Temperature, and Moisture Content
Site Average Correlation Scatterplot

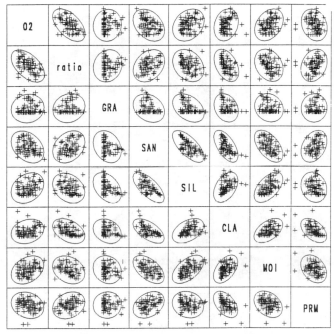

O_2 = log O_2 Rate   Ratio = $(CO_2\ Rate/O_2\ Rate)^{\lambda}$ GRA = Gravel   SAN = Sand   SIL = Silt
CLA = Clay   MOI = Moisture   PRM = log Soil Gas Permeability

	90%	60%	30%	0%
Z1				

Key to Correlation Scatterplots.

Figure A-11.   **Oxygen Utilization Rates, Oxygen:Carbon Dioxide Rate Ratios, Particle Size, Moisture Content, and Soil Gas Permeability Site Average Correlation Scatterplot**

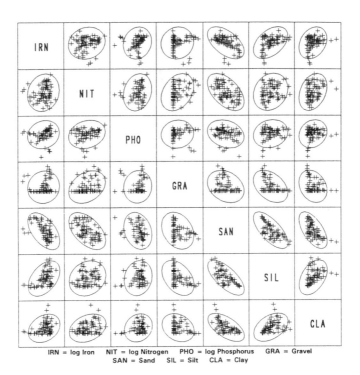

IRN = log Iron   NIT = log Nitrogen   PHO = log Phosphorus   GRA = Gravel
SAN = Sand   SIL = Silt   CLA = Clay

	90%	60%	30%	0%
Z1				

Key to Correlation Scatterplots.

Figure A-12.   **Element Concentrations and Particle Size Site Average Correlation Scatterplot**

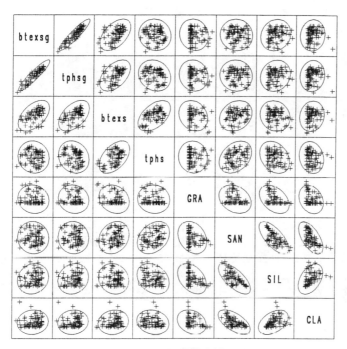

btexsg = log BTEX in Soil Gas   tphsg − log TPH in Soil Gas   btexs = log BTEX Soil
tphs = log TPH in Soil   GRA = Gravel   SAN = Sand   SIL = Silt   CLA = Clay

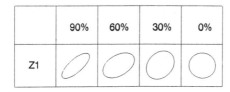

Key to Correlation Scatterplots.

Figure A-13.   Contaminant Concentrations and Particle Size Site Average
                 Correlation Scatterplot

178

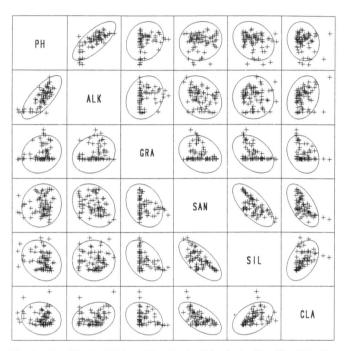

PH = log pH   ALK = log Alkalinity   GRA = Gravel   SAN = Sand   SIL = Silt   CLA = Clay

	90%	60%	30%	0%
Z1				

Key to Correlation Scatterplots.

Figure A-14.   **pH, Alkalinity, and Particle Size Site Average Correlation Scatterplot**

After taking the log transformation, a second regression model was fitted to the oxygen utilization rate using stepwise regression. Finally, the effect of a cyclic seasonal component on the residuals obtained from the fitted regression model was investigated by including the date that the initial in situ respiration test was conducted.

The final regression model for the oxygen utilization rate is:

$$\log\left(O_2\right) = -2.7 + 0.39 \log\left(NIT\right) - 0.108 \left(MOI\right) +$$

$$0.017 \log\left(TPHsg\right) * MOI - 0.004 \log\left(TPHsg\right) * TMP \qquad \text{(A-4)}$$

Each of the effects in the above model is statistically significant at the 0.05 significance level. Note that the effects appearing in the model are consistent with the relationships observed in the bivariate setting. The model explains 41% of the variability in the log-transformed oxygen utilization rate; that is, a 64% correlation between the observed and model-predicted log-transformed oxygen utilization rates. Figure A-15 illustrates actual versus predicted oxygen utilization rates based on model predictions. As shown, the model appears to explain mid-range oxygen utilization rates fairly well, but low oxygen utilization rates are not predicted as accurately. This may be due to an effect on microbial activity not measured during the U.S. Air Force Bioventing Initiative, and therefore, unexplained in the model.

D. CORRELATION OF OXYGEN UTILIZATION AND CARBON DIOXIDE PRODUCTION RATE RATIOS WITH ENVIRONMENTAL PARAMETERS

Because in situ biodegradation rates are measured indirectly through measurements of soil gas oxygen and carbon dioxide concentrations, abiotic processes that affect oxygen and carbon dioxide concentration will affect measured biodegradation rates. The factors that may most influence soil gas oxygen and carbon dioxide concentrations are soil pH, soil alkalinity, and iron content.

At nearly all sites included in the U.S. Air Force Bioventing Initiative, oxygen utilization has proven to be a more useful measure of biodegradation rates than carbon dioxide production. The biodegradation rate in mg of hexane-equivalent/kg of soil per day based on carbon dioxide production usually is less than can be accounted for by the oxygen disappearance. A study conducted at the Tyndall AFB site was an exception. That site had low-alkalinity soils and low-pH quartz sands, and carbon dioxide production actually resulted in a slightly higher estimate of biodegradation (Miller, 1990).

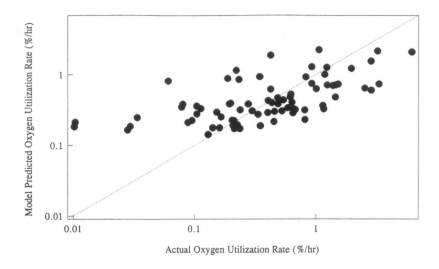

Figure A-15. Actual Versus Model-Predicted Oxygen Utilization Rates

In the case of the higher pH and higher alkalinity soils at sites such as at Fallon NAS and Eielson AFB, little or no gaseous carbon dioxide production was measured (Hinchee et al., 1989; Leeson et al., 1995). This is possibly due to the formation of carbonates from the gaseous evolution of carbon dioxide produced by biodegradation at these sites. A similar phenomenon was encountered by van Eyk and Vreeken (1988) in their attempt to use carbon dioxide evolution to quantify biodegradation associated with soil venting.

In order to determine whether pH and alkalinity influenced carbon dioxide production rates at U.S. Air Force Bioventing Initiative sites, an analysis of the ratio of oxygen utilization to carbon dioxide production versus soil parameters was performed. Due to stoichiometry, the ratio of the oxygen utilization to carbon dioxide production rate will not be 1, because for every 9.5 moles of oxygen consumed, 6 moles of carbon dioxide are produced. A square root transformation of the oxygen utilization and carbon dioxide production rate ratio and log transformations of some of the soil parameters were taken whenever the data were not well represented by the normal distribution. Figures A-9 through A-11 display the bivariate relationships between the ratio and the soil parameters after the transformation. In these figures, as expected, there is a negative correlation between the ratio and the oxygen utilization rate (Figure A-9, r= −0.45). The correlation of the ratio with clay is the most pronounced (Figure A-11, r= −0.40). The ratio is also negatively correlated with pH (Figure A-9, r= −0.25) and alkalinity (Figure A-9, r= −0.30). As noted previously, pH and alkalinity are strongly positively related (Figure A-9, $r = 0.75$). The correlations of the ratio with iron, moisture, permeability, and particle sizes are between 0.20 and 0.30 (Figures A-9 and A-11).

The statistical methods used to model the ratio of the oxygen utilization rate to carbon dioxide production rate as a function of the soil parameters are similar to those used for the oxygen utilization rate analyses. As a preliminary step, a square root transformation of the ratio and log transformation of some of the soil parameters were taken to obtain more normally distributed data. All the transformations for the soil parameters except clay were consistent with those taken previously to model the oxygen utilization rate. A log transformation of clay was considered as it was more correlated with the ratio.

After applying the transformation, a regression model was fitted to the ratio using forward stepwise regression. The model accounted for the effects of each of the soil parameters (except season) and their interactions. Finally, the effect of a cyclic seasonal component on the residuals obtained from the fitted model was determined by incorporating the date the initial in situ respiration test was conducted.

The final model for the ratio of the carbon dioxide production rate to the oxygen utilization rate is as follows:

$$\left(\frac{CO_2\,\text{rate}}{O_2\,\text{rate}}\right)^{\frac{1}{2}} = 1.28 - 0.38\ \log\ (\text{pH}) - 0.095\ \log\ (\text{clay}) + 0.0007\ \log\ (\text{tphs}) * \text{TMP} \qquad \text{(A-5)}$$

Each of the effects in the above model is statistically significant at the 0.05 significance level. The model explains 40% of the variability in the transformed ratio. This amounts to 63% correlation between the observed and model-predicted transformed ratios. The effects of pH on the ratio as predicted by the model are presented in Figure A-16.

The complicated nature of the fitted regression model for the ratio makes the quantification of the effects in the model difficult. However, based on inspection of Figure A-16, it can be seen that as pH increases, the ratio of the carbon dioxide production rate to the oxygen utilization rate decreases, as would be expected given the formation of carbonates.

E.   CORRELATION OF SOIL GAS PERMEABILITY WITH ENVIRONMENTAL PARAMETERS

The bivariate relationships between log-transformed soil gas permeability and each of the independent variables of interest are shown in Figure A-17. In this figure, permeability is most strongly correlated with clay $(r = -0.50)$. The magnitudes of the correlations with both moisture and sand are less pronounced and similar.

The statistical methods used here are similar to those described previously for the oxygen utilization rate and the ratio. Forward stepwise regression was used to determine a regression model for the log-

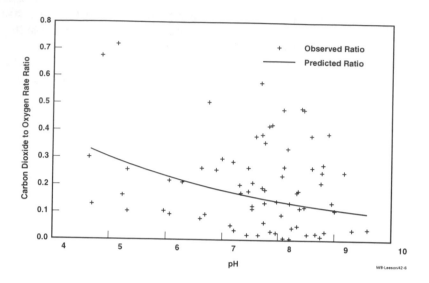

Figure A-16. Variation of pH and the Effect on Oxygen Utilization to Carbon Dioxide Rate Ratio Based on Model Predictions With Average Levels of Other Parameters

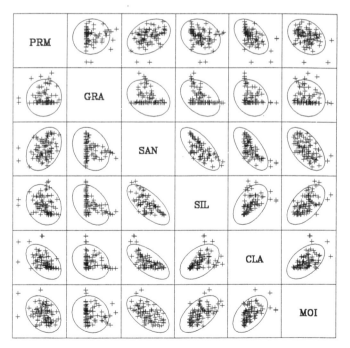

Figure A-17. Soil Gas Permeability, Moisture Content, and Particle Size Site Average Correlation Scatterplot

transformed soil gas permeability. The independent variables of interest in the modeling were moisture content and the particle sizes (sand, silt, and clay).

The final model describing soil gas permeability is given below:

$$\log(\text{PRM}) = 3.2 - 0.064 \text{ clay} \qquad (A\text{-}6)$$

Based on this model, clay alone explains 21% of the variability in the log-transformed soil gas permeability. The effect of clay on soil gas permeability as predicted by the model is presented in Figure A-18. In this figure, the soil gas permeability levels greater than 100 have been censored; that is, they were set to a constant value of 100. Based on the regression model it is determined that an increase in clay by 5 units decreases soil gas permeability by 25% on average.

F. ANALYSES OF DATA FROM CONTAMINATED AND BACKGROUND AREAS

As the preliminary step to comparing the data at background and contaminated sites, transformations of the data parameters were considered. These transformations were consistent with those taken previously to address the other objectives of the statistical analysis. After taking the transformation, statistical analyses were performed separately on each parameter (nitrogen, oxygen utilization rate, etc.). The goal of this analysis was to determine significant differences in the levels of each parameter at background and contaminated sites, with particular interest in TKN concentrations. Measurement of TKN accounts for nitrogen sources within cellular material; therefore, it is possible that TKN concentrations may be higher in contaminated areas, where microbial populations may be higher, than in uncontaminated areas. To date, there is no significant difference between TKN concentrations at contaminated sites (average of 232 mg/kg) and those at background areas (average of 226 mg/kg).

G. SUMMARY

Based on the statistical analyses presented in the previous sections, the following overall conclusions are drawn:

- The relationships between the biodegradation rates and the soil parameters are not very strong. However, some significant relative effects of the soil parameters stand out from the statistical evaluation conducted in the study. Namely, nitrogen, moisture, and soil gas TPH concentrations appear to be the most important characteristics influencing observed field oxygen utilization rates.

184

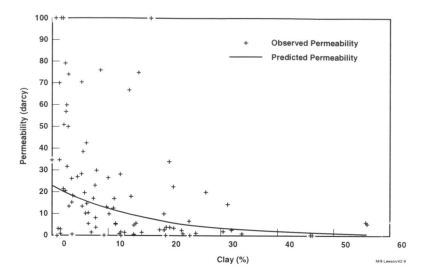

Figure A-18.   Variation of Clay and the Effect on Soil Gas Permeability Based on Model Predictions

- The ratio of the carbon dioxide production rate to the oxygen utilization rate correlates strongly with pH and clay levels in the soil.
- Soil gas permeability correlates with each of the particle sizes (sand, silt, and clay) and moisture content; however, the relative effect of clay on permeability is most important.

The U.S. Air Force Bioventing Initiative has provided a large database of information useful in the design and implementation of bioventing systems. The statistical analyses provide guidelines for determining which parameters are most important to bioventing technology. However, these data must be balanced by experience and site-specific data. For example, sites with relatively low soil nitrogen concentrations should not be discarded as bioventing sites for this reason alone, nor should it be assumed that nitrogen addition at such sites will increase oxygen utilization rates. Data collected from the U.S. Air Force Bioventing Initiative have shown that even sites with low soil nutrient concentrations can exhibit significant microbial activity and would therefore respond well to bioventing.

APPENDIX B

EQUIPMENT SPECIFICATIONS AND MANUFACTURERS

The products and manufacturers listed in this document are intended as guidance for environmental managers and consulting engineers. Products or manufacturers are not endorsed by the U.S. Air Force or the U.S. Environmental Protection Agency.

TABLE OF CONTENTS

I. SOIL GAS SURVEY EQUIPMENT

Calibration Gases

Calibration gases include carbon dioxide, oxygen, and hexane. They are available in the appropriate concentrations for each instrument and may require a special regulator depending on the cylinder type.

The gases are sold through Scott Specialty Gases in Troy, Michigan, (313) 589-2950. The gases cost approximately $124 depending on the cylinder size and gas desired.

185

186

Tedlar™ Sampling Bag

The 1-L bag is made of transparent Tedlar™ and has a polypropylene fitting. The bag is approximately 7×7 inches and is sold in packages of ten. The fitting is opened and closed by twisting the cap, which can also be locked into place.

The Tedlar™ bag is used to store soil gas samples and calibration gases until they can be analyzed by an appropriate gas meter.

The Tedlar™ bags are supplied by SKC, Inc., in Eighty-Four, Pennsylvania, (800) 752-8472. The cost is approximately $82 for 10 bags.

Latex Rubber Tubing

Latex or amber tubing is connected to the Tedlar™ bag tubing fitting for filling the bag. Tubing is normally cut approximately 4 inches in length. Size of tubing is 1/4-inch-O.D. × 3/16-inch-I.D. and can be purchased from VWR Scientific.

Wire/Cable Ties

Nylon cable ties are used like a hose clamp to secure the latex tubing to the tedlar bag fitting. Cable ties can be purchased from Graingers or any hardware store. Catalog # 6X750, pack of 100, $1.91/pack.

Oxygen/Carbon Dioxide Gas Sampling Meter

This hand-held instrument has a rechargeable battery good up to sixteen hours. It has an oxygen and carbon dioxide range of 0 to 25%. The meter has an analog scale readout with audible and visual alarms for low and high warning levels. The meter analyzes oxygen content through an electrochemical cell and carbon dioxide through an infrared sensor. An external filter and an internal filter are employed for high reliability and preventive maintenance. An internal diaphragm pump is provided.

The gas sampling meter is used to determine the oxygen and carbon dioxide content of the ambient air or of the gas within the soil. Calibrations must be performed regularly with gas standards.

The meter is sold by Cascade Associates in Youngstown, Ohio, (216) 758-6649. It costs approximately $3200.

Carrying Case for Gas Sampling Meter

The case is of heavy plastic construction with foam cushioning inside. The case can be secured with locks.

The case is used to protect and carry both the Trace-Techtor and the gas sampling meter.

The case is sold by Cascade Associates in Youngstown, Ohio, (216) 758-6649. It costs approximately $250.

Combustibles Sampling Meter

This meter has a digital display screen with audible and visual alarms for high and low level combustibles/hydrocarbons. They are measured from 0 to 100% LEL and 0 to 10,000 ppm in 20 ppm increments. The meter uses both internal and external filters and includes an internal pump. In addition, it has a data logging function, which permits the meter to be connected with an IBM-compatible computer. It can be operated with alkaline or nicad batteries that hold a 9-hr charge. The platinum catalyst sensor has a flame arrestor.

The meter is used to determine the level of hydrocarbons or combustibles in the ambient air or sampled soil gas. It is a new model which replaces the Trace-Techtor™ meter.

The meter is sold by Cascade Associates in Youngstown, Ohio, (216) 758-6649. It costs approximately $1475.

1:1 Diluter

The diluter is an external fitting that attaches to the inlet of the Trace-Techtor™ meter. It has metal construction and is about 3 inches long. A diluter is required when the oxygen levels of the gas sample drop below twelve percent. At this low oxygen level, the platinum catalyst is not able to combust the gas sample properly.

The function of the 1:1 diluter is to reduce the gas sample flow by one-half. This dilution will reduce the concentration by half. Once a concentration reading is obtained from the meter, it is multiplied by a factor of two to compensate for the dilution.

The diluter is sold by Cascade Associates in Youngstown, Ohio, (216) 758-6649. It costs approximately $150.

10:1 Diluter

This diluter is also an external fitting that attaches to the inlet of the Trace-Techtor™ meter and is small enough to be held in one hand. The diluter has two rotameters built into it to permit a dilution factor up to ten. A diluter is required when the oxygen levels of the gas sample drop below 12%, at which level the platinum catalyst cannot combust the gas sample properly. The 10:1 diluter can be used if the concentration of the sample is still too high to be read after using a 1:1 diluter. This is evident when the gas analysis instrument is pegged at its highest setting.

The function of the 10:1 diluter is to reduce the gas sample flow up to a factor of ten. The dilution factor is set by adjusting the two rotameters until their ratio of the two flows is equal to the dilution ratio. This will reduce the concentration by the same factor. Once a concentration reading is obtained from the meter, it is multiplied by the ratio to compensate for the dilution.

The diluter is sold by Cascade Associates in Youngstown, Ohio, (216) 758-6649. It costs approximately $250.

Trace-Techtor Meter

This hand-held instrument has a rechargeable battery good for ten hours. It is capable of measuring petroleum-based hydrocarbon vapors (BTEX) up to 10000 ppm. It has an analog scale readout with audible and visual alarms for low and high concentration levels. The meter analyzes the vapor through an electrochemical cell with a platinum catalyst. An external filter and an internal filter are employed for high reliability and preventive maintenance. An internal diaphragm pump is also supplied.

The gas sampling meter is used to determine the petroleum hydrocarbon content of the ambient air or of the gas within the soil. Calibrations must be performed regularly with hexane. The instrument can be equipped to detect methane or natural gas.

The meter is sold and manufactured by Gastech in Newark, California, (415) 794-1973. The price is approximately $1500. The Trace-Techtor is no longer manufactured.

Interface Probe

The probe is constructed in the shape of a disk which stores a 100-ft measuring tape and a sensor probe. It weighs sixteen pounds, is $16 \times 18 \times 6$

inches, and is battery-operated. The interface probe resembles a common tape measure but is larger.

The interface probe is very useful when used alone with soil gas probes during site investigation. The probe is used in wells to detect the level at which both oil and water are present. This is accomplished through the use of audible alarms. The probe can detect an oil layer as thin as 0.05 ft.

The interface probe is made by ORS Environmental Systems in Greenville, New Hampshire, (800) 228-2310. It costs approximately $2000.

150-Ft Tape Measure

A 150-ft fiberglass reel tape is needed for site mapping during soil gas survey and also for measuring borehole depths and monitoring point construction. An appropriate measure is available from Graingers catalog #6C192, cost $57.70.

Soil Gas Probes and Well Points (The Macho System)

Electric-powered sampling systems are used for driving soil gas probes. The deluxe system includes a variable-speed hammer drill and the ability to sample soil gas to a depth of approximately ten ft. This is a good starter set, but we would recommend that additional shafts, slotted well points, and hollow probe nipples be purchased. The Macho System costs approximately $3,065 and is available from KVA Analytical Systems, Falmouth MA., (508) 540-0561.

Bulkhead Quick Coupler (Parker)

These brass fittings are threaded into the top of the soil gas probe after it is driven to the desired depth. The fittings allow the sampler an air-tight connection between the probe and the vacuum sampling pump then pulls the soil gas sample from the soil. A supplier is Forberg Scientific, Columbus, Ohio, (614) 294-4600.

Diaphragm Pump (Vacuum/Air Compressor)

The pumps are usually wired for 110 volts for the 1/16-, 1/8-, and 1/3-hp versions. The pumps and compressors are produced by Gast. They are preferred due to their reliability and ease for maintenance.

The pumps are used to draw soil gas from deep monitoring points and soil gas probes. We recommend the 1/3-hp because of the available air produce at 20 psi.

The pump is sold by Grainger in Columbus, Ohio, (800) 323-0620. The costs depend on the size of the pump. A 1/3-hp (catalog #4Z024) costs $228.00.

Probe Puller Adapter

The probe puller adaptor was made by Battelle staff. It is simply a piece of square steel tubing approximately 4 inches by 4 inches by 2 inches wide. A solid probe nipple is then welded in the middle of one outside edge. The adaptor is threaded onto the top of a soil gas probe when sampling is completed. A large utility jack is placed inside the square tube, and the probe is removed.

Utility Jack

The utility jack is used to remove soil gas probes when sampling is completed. It is sold by Graingers (800) 323-0620, Catalog #5Z156, Cost $100.

Miscellaneous Supplies for Soil Gas Survey

Other supplies needed at the site include work gloves, safety glasses, small measuring tape, crescent wrenches, pipe wrenches, vise grips, field record book, cleaning supplies for cleaning soil gas probes, razor blades (single-edge), electrical tape, electrical extension cords, oil, and fuel for generator.

II. VENT WELL INSTALLATION EQUIPMENT

Contracted Drilling Services

If installation of the vent well and soil gas monitoring points is being done by a contracted driller, the driller will provide vent well and well construction materials (sand and bentonite). However, the soil gas monitoring points will need to be furnished to the driller. If no driller is used, the items in this section will need to be acquired.

Hand Augering and Soil Sampling Equipment

A vent well can be installed by hand augering if soil conditions permit. The following is a list of hand augering equipment and equipment needed for collecting soils for laboratory analysis.

Auger Head

It is constructed of stainless steel to resist corrosion and contamination of soil samples. The head is approximately one ft long and is open on both ends to accommodate a soil sample liner. The bottom of the head is flared to allow easy penetration into the ground, and the top has a single bar with a male pipe thread. The male pipe thread attaches to the auger's extension rods.

The auger head is used to house the liner while the soil is being sampled. It is designed to sample the soil with minimal disturbance and effort.

The auger head is supplied by Enviro-Tech Services in Martinez, California, (800) 468-8921. It costs approximately $85.

Core Sampler with Slide Hammer

The core sampler is simply a metal pole with a soil sampler at one end. On the other end is the slide hammer. It is a weight which slides up and down the pole of the core sampler.

The core sampler is another way to obtain undisturbed soil samples. The slide hammer actually drives the sampler into the ground and eliminates the need for the auger head.

The items are supplied by Enviro-Tech Services in Martinez, California, (800) 468-8921. They cost approximately $225.

Sampling Extensions, Extension Cross Handle, Carrying Case

The sampling extensions are long, metal poles which connect the auger head to the cross handle with threaded ends.

The extension cross handle is placed at the top of the auger and used for leverage to turn the auger into the ground. It may have a rubber handle for improved grip.

The metal carrying case is about six ft long and one ft tall and holds the complete auger, disassembled. It has a foam lining to protect contents during travel.

The equipment is supplied by Enviro-Tech Services in Martinez, California, (800) 468-8921. The cost is approximately $400 for all three items.

Brass Sleeves and Plastic End Caps

The sleeve is a cylinder open at both ends and comes in various diameters and lengths. The caps are orange and made of plastic to fit over each end of the sleeve after it is filled with soil.

The sleeve is placed inside the auger head and used as a core sample liner. It contains the soil removed by the auger. The end caps are placed on each end of the sleeve after it is removed from the auger head. Brass sleeves are also used in the core sampler with slide hammer.

The sleeves and caps are supplied by Enviro-Tech Services in Martinez, California, (800) 468-8921. The cost is approximately $3 for both items.

PVC Well Screen

Well screen constructed of PVC is flush-threaded at both ends to accommodate a threaded plug and the riser pipe or blank well casing. Screens are available in 10-, 20-, and 30-slot openings. Well screen is available also in stainless steel.

The screen is sold by Environmental Well Products located in Dayton, Ohio (800) 777-0977. Price varies with size and length

PVC Riser

PVC riser or blank casing also is flush-threaded and has no openings. It is merely an extension of pipe from the well screen to the ground surface. The riser is sold by Environmental Well Products or any drilling supply company.

Bentonite Chips

The chips are available in coarse grades or small pellets. Common sizes include 0.375- and 0.75-inch chips or pellets. They are made from dry bentonite clay and sold in 50-pound bags. The bentonite is chemically stable and able to absorb large amounts of moisture.

The bentonite chips are placed around the necessary equipment within the borehole to form a seal and act as a general filler for the void space. Bentonite was selected because of its high water retention levels. It also interfaces well with Portland cement.

The bentonite is sold by Environmental Well Products Company located in Dayton, Ohio, (800) 777-0977. The price is approximately $10 for 50 pounds.

Silica Sand

The sand contains silica powder for increased chemical stabilization. It is commonly found in the 10×20 graded form.

The silica sand is another form of packing used in well construction. The granular sand is added to boreholes around the screened interval of the vent well and soil gas monitoring points.

The sand is sold by Environmental Well Products Company located in Dayton, Ohio, (800) 777-0977. The price is approximately $6 for 50 pounds.

Concrete Mix

The concrete requires only the addition of water and sets quickly. The concrete is readily available in large quantities throughout the country.

Concrete mix is placed around the manhole at ground level of the well. This ensures its stability during extended absences.

The concrete is sold by Environmental Well Products Company located in Dayton, Ohio, (800) 777-0977. The price is approximately $4 for 50 pounds. It is also available at most building supply stores and hardware stores.

Manhole (Flushmount Well Cover)

Many companies manufacture manholes, some with bolts to secure the top. They are usually sold in 8 inch $\times$ 12 inch or 12 inch $\times$ 12 inch sizes and made of iron, steel, or stainless steel. The bottom is designed to fit over the riser pipe or soil gas monitoring points.

The manhole serves as a marker and gives added protection to the well and the monitoring points.

An appropriate manhole is sold by Environmental Well Products Company located in Dayton, Ohio, (800) 777-0977. The price is approximately $50.

III. SOIL GAS MONITORING POINT EQUIPMENT

Contracted Drilling Services

If installation of the vent well and soil gas monitoring points is being done by a contracted driller, the driller will provide monitoring vent well and well construction materials (sand and bentonite). However the soil gas monitoring points will need to be furnished to the driller. If no driller is used, then items in this section will need to be acquired.

Hand Augering and Soil Sampling

A vent well can be installed by hand augering if soil conditions permit. The following is a list of hand augering equipment and equipment needed for collecting soils for laboratory analysis.

Auger Head

It is constructed of stainless steel to resist corrosion and contamination of soil samples. The head is approximately one ft long and is open on both ends to accommodate a soil sample liner. The bottom of the head is flared to allow easy penetration into the ground, and the top has a single bar with a male pipe thread. The male pipe thread attaches to the auger's extension rods.

The auger head is used to house the liner while the soil is being sampled. It is designed to sample the soil with minimal disturbance and effort.

The auger head is supplied by Enviro-Tech Services in Martinez, California, (800) 468-8921. It costs approximately $85.

Core Sampler with Slide Hammer

The core sampler is simply a metal pole with a soil sampler at one end. On the other end is the slide hammer. It is a weight which slides up and down the pole of the core sampler.

The core sampler is another way to obtain undisturbed soil samples. The slide hammer actually drives the sampler into the ground and eliminates the need for the auger head.

The items are supplied by Enviro-Tech Services in Martinez, California, (800) 468-8921. They cost approximately $225.

Sampling Extensions, Extension Cross Handle, Carrying Case

The sampling extensions are long, metal poles which connect the auger head to the cross handle with threaded ends.

The extension cross handle is placed at the top of the auger and used for leverage to turn the auger into the ground. It may have a rubber handle for improved grip.

The metal carrying case is about six ft long and one ft tall and holds the complete auger, disassembled. It has a foam lining to protect contents during travel.

The equipment is supplied by Enviro-Tech Services in Martinez, California, (800) 468-8921. The cost is approximately $400 for all three items.

Brass Sleeves and Plastic End Caps

The sleeve is a cylinder open at both ends and comes in various diameters and lengths. The caps are orange and made of plastic to fit over each end of the sleeve after it is filled with soil.

The sleeve is placed inside the auger head and used as a core sample liner. It contains the soil removed by the auger. The end caps are placed on each end of the sleeve after it is removed from the auger head. Brass sleeves are also used in the core sampler with slide hammer.

The sleeves and caps are supplied by Enviro-Tech Services in Martinez, California, (800) 468-8921. The cost is approximately $3 for both items.

Suction Strainer

The suction strainer resembles an oxygen diffuser used in fish tanks. It is approximately 0.75 inches in diameter and 8 inches long, constructed of a nylon frame with number 50 mesh screen to permit the flow of gases. The strainers must be tapped with 3/8-inch National Pipe Thread (NPT) in order to install the connector and nylon tubing.

The strainers are filled with aquarium gravel to ensure complete mixture of the soil gas as it is sampled. The strainers are placed at the end of the nylon tubing and set in the monitoring wells, where they are used to withdraw soil gas from the ground, free of dirt and particulate.

The strainer is sold by Grainger in Columbus, Ohio, (800) 323-0620. It costs approximately $7.

NEWLOC™ Male Connector

The male pipe thread connector is made from plastics and has an opening on the end for 0.25-inch tubing. The other end has 0.375-inch male pipe thread.

The connector is used to attach the suction strainer to the nylon tubing in the monitoring wells.

The connector is supplied by New Age Industries, in Willow Grove, Pennsylvania, (215) 657-3151. They cost approximately $1.60 each.

Nylon Tubing

Often called Nylotube, it is made of nylon and sold in various colors for identification purposes. Most common applications of the tube involve the 0.25-inch size.

The tubing transports gases from monitoring points to the surface for soil gas sampling and can be used on some pieces of field equipment for similar purposes. This type of tubing is favored because it is inexpensive, is chemically resistant to hydrocarbons, and is available in many colors. However, the tubing will adsorb some small amount of hydrocarbons.

The tubing is supplied by New Age Industries in Willow Grove, Pennsylvania, (215) 657-3151. It costs approximately $0.36 per ft and is sold in 100-ft rolls.

Quick Connectors (Parker)

Male and female quick connectors and quick connector plugs are compatible with different tube sizes. They are made of brass or stainless steel. The quick connectors offer easy access to monitoring points for taking soil gas samples.

The quick connectors are attached to tubing when quick and convenient access is desired. They also are installed on gas sampling instruments and on tubing found at the monitoring wells. They also give a strong seal to prevent leaking. The quick connector solid plugs are placed in the female quick connectors to prevent corrosion and other forms of damage.

The connectors are sold by Forberg Scientific, located in Columbus, Ohio, (614) 294-4600. A male connector costs approximately $6 and a female approximately $11.

Thermocouple Cable, K Type

The thermocouple cable is a 24-gauge wire insulated with PVC. It can withstand temperatures up to 105°C. It is usually sold by the ft.

The thermocouple is used to measure temperatures, often within a soil gas monitoring point or the outlet stream from a piece of field equipment. The cable transmits the temperature through a current and is recorded using an electronic thermometer.

The cable is supplied by Cole-Parmer in Niles, Illinois, (800) 323-4340. It costs approximately $0.80 per ft.

Thermocouple Minimale Plug

The type K minimale plug has two different prongs and is attached to the thermocouple cable. It acts as a cable termination. It is slightly smaller than a normal electrical plug but serves the same purpose.

The plug is used to connect the thermocouple to the electronic thermometer for collection of temperature data.

The plug is supplied by Cole-Parmer in Niles, Illinois, (800) 323-4340. It costs approximately $5.

Brass Tags

The tags are available in one- to two-inch sizes and in either square or round shape. They are usually constructed of 19-gauge brass. The tags can be purchased with or without labeling.

The tags are stamped, if unlabeled, using a kit and are then placed on wells for identification purposes. They can also be used to label pipes, valves, etc.

The brass tags are manufactured by Seton Identification, New Haven, Connecticut, (800) 754-7360. They are sold in packages of 25 for approximately $20.

Tag Stamping Kit

Stamping kits are sold in sizes from 0.125 to 0.5 inches. They contain both numbers and letters made from steel.

A hammer or mallet is used to stamp the tags with the kit for custom identification.

The stamping kit is manufactured by Seton Identification, New Haven, Connecticut, (800) 754-7360. The kit costs approximately $80.

Bentonite Chips

The chips are available in coarse grades or small pellets. Common sizes include 0.375- and 0.75-inch chips or pellets. They are made from dry bentonite clay and sold in 50-pound bags. The bentonite is chemically stable and able to absorb large amounts of moisture.

The bentonite chips are placed around the necessary equipment within the borehole to form a seal and act as a general filler for the void space. Bentonite was selected because of its high water retention levels. It also interfaces well with Portland cement.

The bentonite is sold by Environmental Well Products Company located in Dayton, Ohio, (800) 777-0977. The price is approximately $10 for 50 pounds.

Silica Sand

The sand contains silica powder for increased chemical stabilization. It is commonly found in the 10×20 graded form.

The silica sand is another form of packing used in well construction. The granular sand is added to boreholes around the screened interval of the vent well and soil gas monitoring points.

The sand is sold by Environmental Well Products Company located in Dayton, Ohio, (800) 777-0977. The price is approximately $6 for 50 pounds.

Concrete Mix

The concrete requires only the addition of water and sets quickly. The concrete is readily available in large quantities throughout the country.

Concrete mix is placed around the manhole at ground level of the well. This ensures its stability during extended absences.

The concrete is sold by Environmental Well Products Company located in Dayton, Ohio, (800) 777-0977. The price is approximately $4 for 50 pounds. It is also available at most building supply stores and hardware stores.

Manhole (Flushmount Well Cover)

Many companies manufacture manholes, some with bolts to secure the top. They are usually sold in 8 inch × 12 inch or 12 inch × 12 inch sizes and made of iron, steel, or stainless steel. The bottom is designed to fit over the riser pipe or soil gas monitoring points.

The manhole serves as a marker and gives added protection to the well and the monitoring points.

An appropriate manhole is sold by Environmental Well Products Company located in Dayton, Ohio, (800) 777-0977. The price is approximately $50.

150-Ft Tape Measure

A 150-ft fiberglass reel tape is needed for site mapping and for measuring borehole depths and during monitoring point sand and bentonite additions. An appropriate tape is sold by Graingers, catalog #6C192, cost $57.70.

Miscellaneous

Cable ties and electrical tape are useful for securing thermocouple wires and nylon tubes together before they are placed in open boreholes.

IV. AIR PERMEABILITY TEST EQUIPMENT

Portable Generator

Several brands are available, and one with a maximum of 5500 watts is recommended. They may be available with wheeled carts. Most have single-phase power available in the two voltage ranges. Most smaller generators run on gasoline, but the larger ones have diesel engines.

A portable generator is essential in a field operation where electrical access is limited. It can power external lighting, pumps, power tools, etc.

The generator is sold by Grainger in Columbus, Ohio, (800) 323-0620. It costs approximately $2200.

Blowers

The blowers recommended are manufactured by Gast. They are oilless regenerative blowers that have a mounted motor. The motors are equipped for different voltage requirements.

The blowers are used during air injection or extraction at a monitoring site. When flammable contamination exists, the blowers should be equipped with explosion-proof circuitry and mufflers.

The blower is sold by Isaacs in Columbus, Ohio, (614) 885-8540. The blower costs vary according to size and power. Example: 2-hp, 145-cfm open flow, cost of $1,100.

Rotameters/Flowmeters

Flowmeters measure the rate at which a gas or liquid is flowing. Rotameters are transparent flowmeters that have the added ability to regulate the flow. The tubes may be constructed of plastic or glass. Each end has a female pipe thread made from brass or plastic. The rotameters are available for various liquid and gas flow levels. Both must be installed in a vertical position for accurate readings.

The rotameter and flowmeter are manufactured and sold by King Instrument Company in Huntington Beach, California, (714) 841-3663. The prices vary depending on which type is needed, but are generally $100 to $200.

Fluke Thermocouple Thermometer

This hand-held, electronic instrument is the size of a large calculator and has a digital readout with an accuracy of 0.1 percent. It operates on a 9-v battery and has two ports for type K, minimale plugs. The thermometer has dual point and differential capability.

The Fluke thermometer is used to record temperature data from the thermocouples. The Fluke thermometer is supplied by several companies including Grainger in Columbus, Ohio, (800) 323-0620. It costs approximately $200.

Dwyer Magnehelic™ Gauges

Magnehelic gauges are used to record negative or positive pressure changes over time during the air permeability test. Four gauges mounted in a panel stand or board should be plumbed in series to cover a wide range of pressures. Appropriate gauges are sold by Graingers, (800) 323-0620, Catalog #3T314, 3T317, 3T319, and 3T321. The cost for each gauge is approximately $51.00.

5-Way Valves (Swagelok®)

The 5-way valve is installed on the magnehelic gauge panel and gives the sampler the ability to record pressures from three points, one after another, simply by turning the valve handle. Sold by Scioto Valve, (614) 891-2617, Part No. B-43ZF2, the valve costs approximately $90.00

Male Non-Valved Quick Couple Plug (Parker®)

Fitting is connected to tubing from 5-way valve. This plug simply plugs into the fitting which is attached to a soil gas monitoring point for measuring pressure during the test. Supplied by Forberg Scientific (614) 294-4600, Part No. 4Z-Q4P-B, Cost $6.00

Stopwatches

A stopwatch is needed by each sampler who is recording pressures at a soil gas monitoring well. Pressures are recorded over time during the air permeability test. Stopwatches can be purchased at most sporting goods stores or at Radio Shack. Cost is about $20.00.

V. IN SITU RESPIRATION TEST EQUIPMENT

Portable Generator

Several brands are available, and one with a maximum of 5500 watts is recommended. They may be available with wheeled carts. Most have single-phase power available in the two voltage ranges. Most smaller generators run on gasoline, but the larger ones have diesel engines.

A portable generator is essential in a field operation where electrical access is limited. It can power external lighting, pumps, power tools, etc.

The generator is sold by Grainger in Columbus, Ohio, (800) 323-0620. It costs approximately $2200.

202

Diaphragm Pump (Vacuum/Air Compressor)

The pumps are usually wired for 110 volts for the 1/16-, 1/8-, and 1/3-hp versions. The pumps and compressors are produced by Gast. They are preferred due to their reliability and ease of maintenance.

The pumps are used to draw soil gas from deep monitoring points and soil gas probes. We recommend the 1/3-hp because of the available air produced at 20 psi.

The pump is sold by Grainger in Columbus, Ohio, (800) 323-0620. The costs depend on the size of the pump, the 1/3-hp (catalog #4Z024) costs $228.00.

Rotameters/Flowmeters

Rotameters are transparent flowmeters with the ability to regulate the flow. The tubes may be constructed of plastic or glass. Each end has a female pipe thread made from brass or plastic. The rotameters will indicate the rate at which the gas is flowing. The flow meter used for in situ respiration testing is connected to the backside of a ⅓-hp diaphragm pump. The flow meter used is normally a 0.4 to 4.0 scfm sold by King Instruments Co. (714) 841-3663. Cost is generally $48.

Helium Leak Detector

The helium leak detector is a rechargeable instrument that can detect helium at concentrations from 0.01 to 100 percent. It operates in a three-stage process, where the sample enters the portable instrument, is analyzed, then purged to the atmosphere. The helium leak detector is approximately 14 inches × 12 inches × 5 inches and weighs seven pounds. The instrument must be calibrated with helium gas.

The helium leak detector is used to detect the presence of helium gas, which is injected into the ground during a tracer test. From this test, an underground model of the gas dispersion can be developed. The detector analyzes soil gas samples from the monitoring wells surrounding the helium injection site.

The leak detector is sold by Mark Products, Inc., in Sunnyvale, California, (800) 621-4600. The price is approximately $4,500.

Compressed Gas Helium 220 ft^3

Helium is mixed with the injection air at approximately 2% helium for the in situ respiration test. Helium can be purchased from compressed gas suppliers or a welding supply. Cost per cylinder is $60.00.

Helium Cylinder Regulator

A two-stage cylinder regulator is necessary for connecting and dispensing the compressed helium gas. The correct connector for cylinder to regulator is a GA 580. Regulators can be purchased through the compressed gas supplier. Cost is approximately $180.00.

Helium/Air Mixing Manifold

The 2% helium mix in air is accomplished by using a one-inch-inside-diameter pipe closed at one end with four tubing connectors which would be plumbed to the diaphragm pumps. The open end of the pipe is where atmospheric air is drawn in for the diaphragm pumps; a tubing connection is installed into the pipe at about six inches from the open end. This connection is for the helium supply to enter the manifold and be swept by incoming air. Helium concentrations need to be measured at the pressure side of the diaphragm pump; if the concentration is too high or low, it can be adjusted at the helium regulator. This item is not commercially available.

Calibration Gases

Calibration gases include helium, carbon dioxide, oxygen, and hexane. They are available in the appropriate concentrations for each instrument and may require a special regulator depending on the cylinder type.

The calibration gases are used to standardize the gas analyzing instruments.

The gases are sold through Scott Specialty Gases in Troy, Michigan, (313) 589-2950. The gases cost approximately $124 depending on the cylinder size and gas desired.

Tedlar™ Sampling Bag

The 1-L bag is made from transparent Tedlar™ and has a polypropylene fitting. The bag is approximately 7 inches × 7 inches and is sold in packages of ten. The fitting is opened and closed by twisting the cap, which can also be locked into place.

The Tedlar™ bag is used to store soil gas samples and calibration gases until they can be analyzed by an appropriate gas meter.

The Tedlar™ bags are supplied by SKC, Inc., in Eighty Four, Pennsylvania, (800) 752-8472. The cost is approximately $82 for 10 bags.

Fluke Thermocouple Thermometer

This hand-held, electronic instrument is the size of a large calculator and has a digital readout with an accuracy of 0.1 percent. It operates on a 9-v battery and has two ports for type K, minimale plugs. The thermometer has dual point and differential capability.

The Fluke thermometer is used to record temperature data from the thermocouples. The Fluke thermometer is supplied by several companies including Grainger in Columbus, Ohio, (800) 323-0620. It costs approximately $200.

Pressure and Vacuum Gauges

Pressure gauges are installed with the flowmeters for air injection. When flow is recorded, the pressure needs to be recorded as well. Vacuum gauges are used on the diaphragm pump used to withdraw soil gas samples from monitoring points and simultaneously record the vacuum. Sold by Graingers (800) 323-0620, the cost is less than $20.00 per gauge, Catalog #1A318.

Oxygen/Carbon Dioxide Gas Sampling Meter

This hand-held instrument has a rechargeable battery good up to 16 hr. It has an oxygen and carbon dioxide range of 0 to 25%. The meter has an analog scale readout with audible and visual alarms for low and high warning levels. The meter analyzes oxygen content through an electrochemical cell and carbon dioxide through an infrared sensor. An external filter and an internal filter are employed for high reliability and preventive maintenance. An internal diaphragm pump is provided.

The gas sampling meter is used to determine the oxygen and carbon dioxide content of the ambient air or of the gas within the soil. Calibrations must be performed regularly with gas standards.

The meter is sold by Cascade Associates in Youngstown, Ohio, (216) 758-6649. It costs approximately $3200.

Carrying Case for Gas Sampling Meter

The case is of heavy plastic construction with foam cushioning inside. The case can be secured with locks,

The case is used to protect and carry both the Trace-Techtor and the gas sampling meter.

The case is sold by Cascade Associates in Youngstown, Ohio, (216) 758-6649. It costs approximately $250.

Combustibles Sampling Meter

This meter has a digital display screen with audible and visual alarms for high and low level combustibles/hydrocarbons. They are measured from 0 to 100% LEL and 0 to 10,000 ppm in 20 ppm increments. The meter uses both internal and external filters and includes an internal pump. In addition, it has a data logging function, which permits the meter to be connected with an IBM-compatible computer. It can be operated with alkaline or nicad batteries that hold a 9-hr charge. The platinum catalyst sensor has a flame arrestor.

The meter is used to determine the level of hydrocarbons or combustibles in the ambient air or sampled soil gas. It is a new model which replaces the Trace-Techtor™ meter.

The meter is sold by Cascade Associates in Youngstown, Ohio, (216) 758-6649. It costs approximately $1475. To find the nearest distributor for Gastech Instruments, call Gastech at (510) 794-6200.

1:1 Diluter

The diluter is an external fitting that attaches to the inlet of the Trace-Techtor™ meter. It has a metal construction and is about three inches long. A diluter is required when the oxygen levels of the gas sample drop below 12%. At this low oxygen level, the platinum catalyst is not able to combust the gas sample properly.

The function of the 1:1 diluter is to reduce the gas sample flow by one-half. This dilution will reduce the concentration by half. Once a concentration reading is obtained from the meter, it is multiplied by a factor of two to compensate for the dilution.

The diluter is sold by Cascade Associates in Youngstown, Ohio, (216) 758-6649. It costs approximately $150.

10:1 Diluter

This diluter is also an external fitting that attaches to the inlet of the Trace-Techtor™ meter and is small enough to be held in one hand. The diluter has two rotameters built into it to permit a dilution factor up to ten. A diluter is required when the oxygen levels of the gas sample drop below 12%, at which levels the platinum catalyst cannot combust the gas sample properly. The 10:1 diluter can be used if the concentration of the sample is still too high to be read after using a 1:1 diluter. This is evident when the gas analysis instrument is pegged at its highest setting.

The function of the 10:1 diluter is to reduce the gas sample flow up to a factor of ten. The dilution factor is set by adjusting the two rotameters until their ratio of the two flows is equal to the dilution ratio. This will reduce the concentration by the same factor. Once a concentration reading is obtained from the meter, it is multiplied by the ratio to compensate for the dilution.

The diluter is sold by Cascade Associates in Youngstown, Ohio, (216) 758-6649. It costs approximately $250.

VI. MISCELLANEOUS ITEMS

Teflon™ Thread Tape

The white tape is made of Teflon™ and comes in rolls of 0.25-, 0.5- and 1-inch widths. The tape is wrapped over pipe threading to prevent leaking of liquids and gases. The tape is supplied by U.S. Plastics Corporation in Lima, Ohio, (800) 357-9724. It costs approximately $1.

PVC Piping Supplies

PVC pipe is needed in various diameters up to 6 inches. Most piping used is schedule 40 and in 10- or 20-ft lengths. Some of the supplies (e.g., valves, tees, and couplings) may be needed as schedule 80 PVC.

The PVC piping is used to transport gases (usually air), to vent wells, or to transport liquids from contaminated wells.

The items are supplied by U.S. Plastics Corporation in Lima, Ohio, (800) 357-9724. Costs depend on the specific piping size, length, and schedule required.

PVC Pipe Cement and Primer

The PVC primer is a volatile, clear liquid that is applied with a small sponge. The PVC cement is a viscous and gray liquid also applied with a sponge. Both have a strong odor and can be harmful if used without proper ventilation.

The primer is used to clean and prime the PVC before assembly. After the primer dries, the cement is applied to connect the PVC pieces. The PVC cement sets quickly.

The items are supplied by U.S. Plastics Corporation in Lima, Ohio, (800) 357-9724. The cost is approximately $20 for both the cement and primer.

Pipe Fittings

Many different types and sizes of pipe fitting are needed for pump connections and tubing connections. The Graingers catalog shows a large selection of reasonably priced steel and brass pipe fittings.

VII. OPTIONAL ITEMS

Soil Moisture Meter

The soil moisture meter is an electronic, hand-held instrument that operates from a 9-v battery. Two spring terminals at the top of the meter are used to connect the moisture blocks.

The meter gives a digital display of the soil moisture content as a percentage obtained from the soil moisture blocks.

The meter is supplied by Soilmoisture Equipment Corporation in Santa Barbara, California, (805) 964-3525. It costs approximately $310.

Soil Moisture Blocks

The blocks consist of a lead wire connected to the gypsum block, which is a 1-inch-diameter cylinder. The blocks have a life expectancy of 3 to 5 yrs. The gypsum is able to compensate for varying salinity conditions.

They are placed in the soil to transmit the soil moisture content to the soil moisture meter using an electric current. They are available in different lengths and are installed along with the soil gas monitoring points.

The block is supplied by Soilmoisture Equipment Corporation in Santa Barbara, California, (805) 964-3525. It costs approximately $15.

Bailer

Constructed of Teflon™, PVC, or stainless steel, the bailers are available in 1- to 4-ft lengths. Teflon™ is preferred for its chemically inert properties and low cost.

The bailers are lowered into the wells by cords or rope to remove water or other standing liquids. The well must be dry to install the screens and suction strainers. No soil gas sampling can occur due if liquid(s) are present.

The bailer is sold by Environmental Well Products Company, located in Dayton, Ohio, (800) 777-0977. The price is approximately $140.

APPENDIX C

EXAMPLE PROCEDURES FOR CONDUCTING
BIOVENTING TREATABILITY STUDIES

I. EXAMPLE PROCEDURE FOR COLLECTION, LABELING, PACKING, AND SHIPPING OF SOIL SAMPLES

A. SCOPE/PURPOSE

This procedure describes collection, labeling, packing, and shipping of soil samples as well as decontamination procedures for sampling equipment.

B. DEFINITIONS

Sampling Team: People responsible for collecting and processing soil samples.

C. PROCEDURE

1. Sample Collection

Soil samples usually are collected with split-spoon samplers during soil-boring operations or with hand-held soil augers. Regardless of how samples are collected, all equipment will be decontaminated prior to and after collection of each sample.

a. Equipment Decontamination
- The sampler will be washed thoroughly.
- Rinsed with deionized or distilled water.
- Rinsed with methanol and allowed to air dry.
- Rinsates will be disposed of in an environmentally sound manner.

b. Sample Collection
- At a minimum, rubber or vinyl gloves will be worn to collect the sample. If higher levels of contamination are anticipated, nitrile or nitrocellulose gloves will be worn in addition to other appropriate safety gear as indicated in the Site Health and Safety Plan.
- During processing of soil samples, the work area should be covered with vinyl or plastic. Between samples, the work area should be cleaned of soil residues. The work area should be positioned upwind of the test area or drill rig.

209

210

- For split-spoon sampling, the soil core usually is retained in the stainless steel or brass sampling tube. The tube should be capped top and bottom after a Teflon™ liner or its equivalent has been placed over the exposed soil.
- If the soil is to be transferred to other containers, such as those listed below for various analysis types, scoop the sample directly into the sample container. If organic analyses are to be performed, the scoop should be stainless steel. A soil core sample will be spooned or scooped directly from the coring tube, split spoon, etc. into the sample container.
- If a gloved hand comes into contact with the sample, then new gloves will be used for each sample. In addition, a background sample that contacts a glove will be collected as a control.

c. Split Samples

A homogenous mix for a split soil sample can be obtained by mixing soil in a stainless steel pan and filling both sample containers with alternate spoonfuls. However, if a sample is collected for trace volatile analysis, too much sample agitation and mixing can drive off the compounds of concern. Consequently, if a split-spoon or other soil sample for volatile organic analysis is to be split and there is concern that the above homogenization would cause trace volatile compounds to be lost, an alternate splitting technique will be used. The undisturbed core or soil will be spooned directly into the two jars by alternating spoonfuls between the sample and the split container. This will ensure a fairly even split while reducing the agitation and exposure of the sample surface area.

d. Sample Containers and Sample size

Soil samples will be stored in appropriate containers as indicated in the site test plan or as directed by the analytical laboratory. For sample size requirements, refer to the site test plan or discuss with analytical laboratory. Some suggested container types and sample sizes are as follows:

- Volatile analysis: Glass jar, wide mouth, Teflon-faced cap, 125-mL capacity, 100 g sample volume minimum
- Semi-volatiles: Glass jar, wide mouth, Teflon-faced cap, 125-mL capacity, 100 g sample volume minimum
- Metals: HDPE or glass wide-mouth jar

- For other soil analysis types including particle-size analysis, nutrient analysis, and moisture determination, samples can be stored in metal, plastic, or glass containers.

2. Sample Label and Log

A sample must be labeled with enough information for all parties who may have to deal with it. Refer to the test/project plan for labeling instruction. At a minimum, the samples are to be labeled with the following information:

- Test Site where sample was collected
- Soil Boring Number or ID
- Soil sampling depth
- Initials of sampler
- Date and time of collection
- Information to be recorded in the log/record book, including specific equipment used, sampler, date and time, and any observations about the sampled material or meter readings taken.

3. Sample Packing and Shipping

- The soil samples will be placed in plastic bags and stored in a refrigerator, ice chest, or insulated box on ice immediately after being placed in appropriate containers and labeled. Be sure sample containers and bags are tightly closed and that there is sufficient ice to maintain refrigerated conditions until samples arrive at the laboratory.
- Control samples and field blanks will not be shipped with contaminated samples.
- Complete chain-of-custody forms for each cooler. Refer to SOP ENVIR. I-005-00 for proper procedure for completion of chain-of-custody documentation.
- Ship samples to arrive within 24 hr whenever possible. Shipment will be made by Federal Express (when possible) — Priority Overnight Service with Saturday deliveries specified when applicable.
- Notify recipient about specifics of shipment.

4. Quality Control

- Descriptions and dates of all of the above activities will be documented in study records.
- Soil analysis information will be included in the study records. Photographs will be taken periodically and retained with the study records.

- Records will be kept as indicated in this procedure and will be reviewed periodically by the study/task leader.

II. EXAMPLE PROCEDURE FOR IN SITU RESPIRATION TESTING

A. SCOPE/PURPOSE

This section describes procedures for conducting an in situ respiration test.

B. REFERENCES

Hinchee, R.E. and S.K. Ong. 1992. "A Rapid In Situ Respiration Test for Measuring Aerobic Biodegradation Rates of Hydrocarbons in Soil." *Journal of Air & Waste Management Association,* 42(10):1305-1312.

Hinchee, R.E., S.K. Ong, R.N. Miller, D.C. Downey, and R. Frandt. 1992. *Test Plan and Technical Protocol for a Field Treatability Test for Bioventing*, Rev. 2. U.S. Air Force Center for Environmental Excellence, Brooks Air Force Base, TX.

C. DEFINITIONS

Sampling Team: People responsible for conducting the in situ respiration test.

D. PROCEDURE
1. Field Instrumentation and Measurement
a. Oxygen and Carbon Dioxide

Gaseous concentrations of carbon dioxide and oxygen will be analyzed using a GasTech model 3252OX carbon dioxide/oxygen analyzer or equivalent. The battery charge level will be checked to ensure proper operation. The air filters will be checked and, if necessary, cleaned or replaced before the experiment is started. The instrument will be turned on and equilibrated for at least 30 minutes before conducting calibration or obtaining measurements. The sampling pump of the instrument will be checked to ensure that it is functioning. Low flow of the sampling pump can indicate that the battery level is low or that some fines are trapped in the pump or tubing.

Before use each day, meters will be calibrated against purchased carbon dioxide and oxygen calibration standards. These standards will be selected to be in the concentration range of the soil gas to be sampled. The carbon dioxide calibration will be performed against atmospheric carbon dioxide (0.05%) and a 5% standard. The oxygen will be calibrated using atmospheric oxygen (20.9%) and against a 5% and 0% standard. Standard gases will be purchased from a specialty gas supplier. To calibrate the instrument with standard gases, a Tedlar™ bag (capacity ~1 L) is filled with the standard gas, and the valve on the bag is closed. The inlet nozzle of the instrument is connected to the Tedlar™ bag, and the valve on the bag is opened. The instrument is then calibrated against the standard gas according to the manufacturer's instructions. Next, the inlet nozzle of the instrument is disconnected from the Tedlar™ bag, and the valve on the bag is shut off. The instrument will be rechecked against atmospheric concentration. If recalibration is required, the above steps will be repeated.

b. Hydrocarbon Concentration

Petroleum hydrocarbon concentrations will be analyzed using a GasTech Trace-Techtor™ hydrocarbon analyzer (or equivalent) with range settings of 100 ppm, 1,000 ppm, and 10,000 ppm. The analyzer will be calibrated against two hexane calibration gases (500 ppm and 4,400 ppm). The Trace-Techtor™ has a dilution fitting that can be used to calibrate the instrument in the low-concentration range.

Calibration of the GasTech Trace-Techtor™ is similar to the GasTech Model 32402X, except that a mylar bag is used instead of a Tedlar™ bag. The oxygen concentration must be above 10% for the Trace-Techtor™ analyzer to be accurate. When the oxygen drops below 10%, a dilution fitting must be added to provide adequate oxygen for analysis.

Hydrocarbon concentrations can be determined also with a flame ionization detector (FID), which can detect low (below 100 ppm) concentrations. A photoionization detector (PID) is *not* acceptable.

c. Helium Monitoring

Helium in the soil gas will be measured with a Marks Helium Detector Model 9821 or equivalent with a minimum sensitivity of 100 ppm (0.01%). Calibration of the helium detector follows the same basic procedure described for oxygen calibration, except that the setup for calibration is different. Helium standards used are 100 ppm (0.01%), 5,000 ppm (0.5%), and 10,000 ppm (1%).

d. Temperature Monitoring

In situ soil temperature will be monitored using Omega Type J or K thermocouples (or equivalent). The thermocouples will be connected to an Omega OM-400 Thermocouple Thermometer (or equivalent). Each thermocouple will be calibrated against ice water and boiling water by the contractor before field installation.

e. Airflow Measurement

Before respiration tests are initiated at individual monitoring points, air will be pumped into each monitoring point using a small air compressor, as described in Section 5.7. Airflow rates of 1 to 1.5 cfm will be used, and flow will be measured using a Cole-Palmer Variable Area Flowmeter No. N03291-4 (or equivalent). Helium will be introduced into the injected air at a 1% concentration. A helium flow rate of approximately 0.01 to 0.015 cfm (0.6 to 1.0 cfh) will be required to achieve this concentration. A Cole-Palmer Model L-03291-00 flowmeter or equivalent will be used to measure the flow rate of the helium feed stream.

2. In Situ Respiration Test Procedures

The in situ respiration test should be conducted using at least four screened intervals of the monitoring points and a background well. The results from this test will determine if in situ microbial activity is occurring and if it is oxygen-limited.

a. Test Implementation

Air with 1 to 2% helium will be injected into the monitoring points and background well. Following injection, the change of oxygen, carbon dioxide, total hydrocarbon, and helium in the soil gas will be measured over time. Helium will be used as an inert tracer gas to assess the extent of diffusion of soil gases within the aerated zone. If the background well is screened over an interval greater than 10 ft, the required air injection rate may be too high to allow helium injection. The background monitoring point will be used to monitor natural degradation of organic matter in the soil.

Oxygen, carbon dioxide, and total hydrocarbon levels will be measured at the monitoring points before air injection. Normally, air will be injected into the ground for at least 20 hr at rates ranging from 1.0 to 1.7 cfm (60 to 100 cfh). The blowers used will be diaphragm compressors Model 4Z024 from Grainger (or equivalent) with a nominal capacity of 1.7 cfm (100 cfh) at 10 psi. The helium used as a tracer will be 99% or greater purity, which is available from most welding supply stores. The flow rate of helium will be adjusted to 0.6 to 1.0 cfh to obtain about 1% in the final air mixture which will be injected into the contaminated area. Helium in the soil gas will be measured with a Marks Helium Detector Model 9821 (or equivalent) with a minimum sensitivity of 0.01%.

After air and helium injection is completed, the soil gas will be measured for oxygen, carbon dioxide, helium, and total hydrocarbon. Soil gas will be extracted from the contaminated area with a soil gas sampling pump system. Typically, measurement of the soil gas will be conducted at 2, 4, 6, and 8 hours and then every 4 to 12 hours, depending on the rate at which the oxygen is utilized. If oxygen uptake is rapid, more frequent monitoring will be required. If it is slower, less frequent readings will be acceptable.

At shallow monitoring points, there is a risk of pulling in atmospheric air in the process of purging and sampling. Excessive purging and sampling may result in erroneous readings. There is no benefit in over sampling, and when sampling shallow points, care will be taken to minimize the volume of air extraction. In these cases, a low-flow extraction pump of about 0.03 to 0.07 cfm (2.0 to 4.0 cfh) will be used. Field judgment will be required at each site in determining the sampling frequency.

The in situ respiration test will be terminated when the oxygen level is about 5%, or after 5 days of sampling. The temperature of the soil before air injection and after the in situ respiration test will be recorded.

b. Data Interpretation
Data from the in situ respiration tests will be summarized, and their oxygen utilization rates computed. Details on data interpretation are presented in Chapter 5.

3. Quality Control

- Descriptions and dates of all of the above activities will be documented in study records.
- Soil analysis information will be included in the study records. Photographs will be taken periodically and retained with the study records.
- Records will be kept as indicated in this procedure and will be periodically reviewed by the study/task leader.

III. EXAMPLE PROCEDURE FOR SOIL GAS PERMEABILITY TESTING

A. SCOPE/PURPOSE
This section describes procedures for conducting a soil gas permeability test.

B. DEFINITIONS
Sampling Team: People responsible for conducting the soil gas permeability test.

C. PROCEDURE
1. Field Instrumentation and Measurement
a. Oxygen and Carbon Dioxide
Gaseous concentrations of carbon dioxide and oxygen will be analyzed using a GasTech model 3252OX carbon dioxide/oxygen analyzer or equivalent. The battery charge level will be checked to ensure proper operation. The air filters will be checked and, if necessary, cleaned or replaced before the experiment is started. The instrument will be turned on

and equilibrated for at least 30 minutes before conducting calibration or obtaining measurements. The sampling pump of the instrument will be checked to ensure that it is functioning. Low flow of the sampling pump can indicate that the battery level is low or that some fines are trapped in the pump or tubing.

Meters will be calibrated each day prior to use against purchased carbon dioxide and oxygen calibration standards. These standards will be selected to be in the concentration range of the soil gas to be sampled. The carbon dioxide calibration will be performed against atmospheric carbon dioxide (0.05%) and a 5% standard. The oxygen will be calibrated using atmospheric oxygen (20.9%) and against a 5% and 0% standard. Standard gases will be purchased from a specialty gas supplier. To calibrate the instrument with standard gases, a Tedlar™ bag (capacity ~1 L) is filled with the standard gas, and the valve on the bag is closed. The inlet nozzle of the instrument is connected to the Tedlar™ bag, and the valve on the bag is opened. The instrument is then calibrated against the standard gas according to the manufacturer's instructions. Next, the inlet nozzle of the instrument is disconnected from the Tedlar™ bag and the valve on the bag is shut off. The instrument will be rechecked against atmospheric concentration. If recalibration is required, the above steps will be repeated.

b. Hydrocarbon Concentration

Petroleum hydrocarbon concentrations will be analyzed using a GasTech Trace-Techtor™ hydrocarbon analyzer (or equivalent) with range settings of 100 ppm, 1,000 ppm, and 10,000 ppm. The analyzer will be calibrated against two hexane calibration gases (500 ppm and 4,400 ppm). The Trace-Techtor™ has a dilution fitting that can be used to calibrate the instrument in the low-concentration range.

Calibration of the GasTech Trace-Techtor™ is similar to the GasTech Model 32402X, except that a mylar bag is used instead of a Tedlar™ bag. The oxygen concentration must be above 10% for the Trace-Techtor™ analyzer to be accurate. When the oxygen drops below 10%, a dilution fitting must be added to provide adequate oxygen for analysis.

Hydrocarbon concentrations can also be determined with a flame ionization detector (FID), which can detect low (below 100 ppm) concentrations. A photoionization detector (PID) is *not* acceptable.

c. Pressure/Vacuum Monitoring

Changes in soil gas pressure during the air permeability test will be measured at monitoring points using Magnehelic™ or equivalent gauges. Tygon™ or equivalent tubing will be used to connect the pressure/vacuum gauge to the quick-disconnect on the top of each monitoring point. Similar gauges will be positioned before and after the blower unit to measure pressure at the blower and at the head of the venting well. Pressure gauges are available in a variety of pressure ranges, and the same gauge can be used to measure either positive or negative (vacuum) pressure by simply switching inlet ports. Gauges are sealed and calibrated at the factory and will be rezeroed before each test. The following pressure ranges (in inches H_2O) will typically be available for this field test: 0-1", 0-5", 0-10", 0-20", 0-50", 0-100", and 0-200".

Air pressure during injection for the in situ respiration test will be measured with a pressure gauge having a minimum range of 0 to 30 psig.

d. Airflow Measurement

During the air permeability test, an accurate estimate of flow (Q) entering or exiting the vent well is required to determine k and R_I. Several airflow measuring devices are acceptable for this test procedure.

Pitot tubes or orifice plates combined with an inclined manometer or differential pressure gauge are acceptable for measuring flow velocities of 1,000 ft/min or greater (~ 20 scfm in a 2-in. pipe). For lower flow rates, a large rotameter will provide a more accurate measurement. If an inclined manometer is used, the manometer must be rezeroed before and after the test to account for thermal expansion/contraction of the water. Devices to measure static and dynamic pressure must also be installed in straight pipe sections according to manufacturer's specifications. All flow rates will be corrected to standard temperature and ambient pressure (altitude) conditions.

2. Soil Gas Permeability Test Procedures

This section describes the field procedures that will be used to gather data to determine k and to estimate R_I.

Before the soil gas permeability test is initiated, the site will be examined for any wells (or other structures) that will not be used in the test but may serve as vertical conduits for gas flow. These will be sealed to prevent short-circuiting and to ensure the validity of the soil gas permeability test.

a. System Check

Before proceeding with this test, soil gas samples will be collected from the vent well, the background well, and all monitoring points and analyzed for oxygen, carbon dioxide, and volatile hydrocarbons. After the blower system has been connected to the vent well and the power has been hooked up, a brief system check will be performed to ensure proper operation of the blower and the pressure and airflow gauges and to measure an initial pressure response at each monitoring point. This test is essential to ensure that the proper range of Magnehelic™ gauges are available for each monitoring point at the onset of the soil gas permeability test. Generally, a

10- to 15-minute period of air extraction or injection will be sufficient to predict the magnitude of the pressure response, and the ability of the blower to influence the test volume.

b. Soil Gas Permeability Test

After the system check, and when all monitoring point pressures have returned to zero, the soil gas permeability test will begin. Two people will be required during the initial hour of this test — one to read the Magnehelic™ gauges and the other to record pressure (P') versus time on the data sheet. This will improve the consistency in reading the gauges and will reduce confusion. Typically, the following test sequence will be followed:

1. Connect the Magnehelic™ gauges to the top of each monitoring point with the stopcock opened. Return the gauges to zero.
2. Turn the blower unit on, and record the starting time to the nearest second.
3. At 1-minute intervals, record the pressure at each monitoring point beginning at $t = 60$ s.
4. After 10 minutes, extend the interval to 2 minutes. Return to the blower unit and record the pressure reading at the well head, the temperature readings, and the flow rate from the vent well.
5. After 20 minutes, measure P' at each monitoring point in 3-minute intervals. Continue to record all blower data at 3-minute intervals during the first hour of the test.
6. Continue to record monitoring point pressure data at 3-minute intervals until the 3-minute change in P' is less than 0.1 in. of H_2O. At this time, a 5- to 20-minute interval can be used. Review data to ensure accurate data were collected during the first 20 minutes. If the quality of these data is in question, turn off the blower, allow all monitoring points to return to zero pressure, and restart the test.
7. Begin to measure pressure at any groundwater monitoring points that have been converted to monitoring points. Record all readings, including zero readings and the time of the measurement. Record all blower data at 30-minute intervals.
8. Once the interval of pressure data collection has increased, collect soil gas samples from monitoring points and the blower exhaust (if extraction system), and analyze for oxygen, carbon dioxide, and hydrocarbons. Continue to gather pressure data for 4 to 8 hours. The test will normally be continued until the outermost monitoring point with a pres-

sure reading does not increase by more than 10% over a 1-hour interval.

9.   Calculate the values of k and R_I with the data from the completed test; the HyperVentilate™ computer program is recommended for this calculation.

c. Post-Permeability Test Soil Gas Monitoring

Immediately after the permeability test is completed, soil gas samples will be collected from the vent well, the background well, and all monitoring points, and analyzed for oxygen, carbon dioxide, and hydrocarbons. If the oxygen concentration in the vent well has increased by 5% or more, oxygen and carbon dioxide will be monitored in the vent well in a manner similar to that described for the monitoring points in the in situ respiration test. (Initial monitoring may be less frequent.) The monitoring will provide additional in situ respiration data for the site.

3. Quality Control
- Descriptions and dates of all of the above activities will be documented in study records.
- Soil analysis information will be included in the study records. Photographs should be taken periodically and retained with the study records.
- Records will be kept as indicated in this procedure and will be reviewed periodically by the study/task leader.

APPENDIX D

OFF-GAS TREATMENT OPTIONS

Off-gas treatment is not typically a component of bioventing systems. Bioventing systems are usually configured to inject air into the in situ soil mass. The injected air then moves through the soil to act as an oxygen source for microbial activity. The bioventing injection air flow rate is low and is selected to minimize discharge from the surface while providing an adequate supply of oxygen for the organisms.

Air injection is the preferred bioventing configuration; however, air extraction may be necessary at sites where movement of vapors into subsurface structures or air emissions are difficult to control. If a building or other structure is located within the radius of influence of a site, or if the site is near a property boundary beyond which hydrocarbon vapors cannot be pushed, air extraction may be considered. A significant disadvantage of the air extraction configuration is that biodegradation is limited to the contaminated soil because vapors do not move outward, creating an expanded bioreactor. The result is less biodegradation and more volatilization. In general, increasing extraction rates will increase both volatilization and biodegradation rates until the site becomes aerated. At this point, increasing flow rate will not increase biodegradation, but will continue to increase volatilization. The optimal input air flow is the minimum extraction rate that satisfies the oxygen demand. Some volatilization will occur regardless of the extraction rate. The relative removal attributed to biodegradation and volatilization is quite variable and site-dependent. At a JP-4 jet-fuel-contaminated site at Tyndall AFB, Miller et al. (1991) found that it was possible at the optimal air flow rate to achieve about 85% contaminant removal by biodegradation.

Currently, only 6 of 125 Bioventing Initiative sites use air extraction to oxygenate the site. Two of the sites — Davis Global Communications Site (near McClellan AFB) and BX Service Station, Patrick AFB — operated in extraction mode for 60 to 90 days, at which time the system was reconfigured for air injection because vapor concentrations had been significantly reduced. At Patrick AFB, initial vapor concentrations of total petroleum hydrocarbons (TPH) were as high as 27,000 ppmv. After approximately 75 days of operation, concentrations were reduced to 1,600 ppmv, and the bioventing system was reconfigured for injection. An additional site is the Base Service Station at Vandenberg AFB. Because this site contains high concentrations of the more volatile components of

221

222

fuels and is an active service station, vapors could migrate into the building on site. This bioventing system was operated in an extraction configuration in two Phases (Downey et al., 1994). During Phase I, extracted soil gas was passed through a PADRE® vapor treatment system, where high concentrations of volatiles were adsorbed and condensed to liquid fuel. The treated soil gas then was recirculated through the soil using air injected via biofilter trenches located along the perimeter of the site. Phase II was initiated once TVH concentrations were reduced to <1,000 ppmv. At this time, the PADRE® system was taken off line, and the extracted soil gas was reinjected directly into the biofilter trenches.

This section discusses minimization of the off-gas flow rate, seven commercially available alternatives for treating organic vapors in an air stream, and some emerging vapor treatment technologies. The vapor treatment technology discussions in this section are derived from information on remedial technologies published by AFCEE (1992, 1994) and a description of off-gas treatment in a Reference Handbook for soil vapor extraction technology (U.S. EPA, 1991, EPA/540/2-91/003). Figure D-1 shows the general ranges of applicability for some commonly used off-gas treatment methods. The organic vapor treatment options discussed in the following sections are

- limiting off-gas production
- direct discharge
- off-gas reinjection
- biofiltration
- adsorption on carbon or resin
- catalytic oxidation
- flame incineration
- thermal destruction in an internal combustion engine
- emerging vapor treatment technologies

Many of these methods have been used in industrial applications to control point source VOC emissions. Figure D-1 shows that most of these alternatives may be used over a range of concentrations that spans several orders of magnitude. Usually, however, each option is cost-effective over a small part of that range. For example, granular activated carbon (GAC) adsorption could be used to treat a vapor stream containing 10,000 ppmv of hydrocarbon vapors, but the cost for carbon regeneration would be prohibitive.

As shown in Figure D-1, thermal treatment methods are more cost-effective for treating off-gas containing higher concentrations of vapor contaminants. No distinct guidelines exist for selecting thermal treatment units for specific applications, but the tradeoff between capital and operating cost sets general ranges of applicability for the thermal treatment

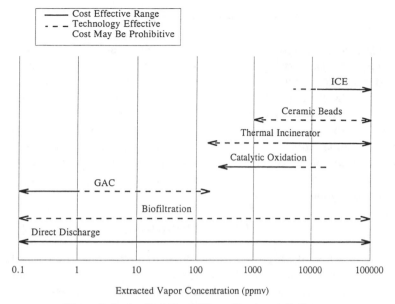

Figure D-1. Applicability of Vapor Treatment Options

methods. Catalytic oxidation units usually have higher initial cost but lower fuel requirements than flame incinerators. As a result, the catalytic oxidation units usually are economical for influent containing less than 5,000 ppmv of contaminants. The capital cost of internal combustion engine (ICE) treatment units is similar to catalytic oxidation units. The ICE is not limited to operating with an inlet combustible vapor concentration below 25% of the lower explosive limit (LEL). The ICE units, therefore, gain a significant advantage when the vapor concentration is over 25% of LEL.

Limiting Off-Gas Production

Design and operating features can minimize the volume of off-gas released by bioventing systems. This source reduction approach to pollution prevention should be used whenever possible at bioventing sites. Options for minimizing off-gas production include using the lowest air flow rate possible while still supplying sufficient air and/or using air injection instead of air extraction configurations to aerate the contaminated area. Bioventing systems can be operated at much lower air flow rates than standard soil vapor extraction systems. A well designed and operated bioventing system can minimize off-gas releases without compromising the oxygenation of the contaminated area. As discussed above, air injection systems are preferred unless site conditions require air extraction to control movement or accumulation of contaminant vapors.

Direct Discharge

Direct discharge involves release of air containing organic vapors directly through a stack. The stack will disperse the vapors but not remove or destroy contaminants. When the organic vapor concentration in the extraction well off-gas stream is low or in localities with less stringent air treatment standards, treatment may not be required. Direct discharge of vapors to the atmosphere can be a viable option when consistent with good environmental practice and local permitting requirements. The concentration of the contaminants, the off-gas release rate, and the location and type of nearby receptors are considered when evaluating direct discharge options.

Off-Gas Reinjection

Reinjection of off-gas for further biodegradation can be a cost-effective and environmentally sound treatment option. Off-gas reinjection configurations offer the advantages of low surface emissions and no point source generation. The reinjection treatment option consists of distributing extracted air with contaminant vapors back into the soil to allow in situ aerobic biodegradation to destroy the contaminants. Reinjection is accomplished by piping the discharge of the extraction blowers to air distribution wells or trenches where the air infiltrates back into the soil. In situ respiration and soil gas permeability data must be available for the site. These data indicate the expected biodegradation rate and radius of influence, which are needed to determine the design capacity for the reinjection point. The soil volume available must be sufficient to accept the off-gas air flow and allow biodegradation of the contaminant mass flow in the off-gas.

Reinjection wells should be located and designed to ensure that the reinjection process is destroying the contaminants rather than increasing contaminant migration. After reinjection is established, surface emission testing may be performed to ensure that contaminants are not escaping at the site surface. Soil gas monitoring should be performed to ensure that contaminant migration is not being increased. Monitoring of migration is particularly important at sites where air extraction is necessary because buildings are present.

Biofiltration

Biofiltration can be used to destroy a variety of volatile organic contaminants in an off-gas stream. The biofiltration process uses microorganisms immobilized as a biofilm on a porous filter substrate such as peat or compost. As the air and vapor contaminants pass through the filter, contaminants transfer from the gas phase to the biolayer where they are metabolized. Influent contaminant concentrations less than about 1,000 ppmv can be treated with a typical contact time of 15 to 90 seconds (Skladany, et al., 1994). Vendor data indicate that biofiltration is most effective for gasoline hydrocarbon vapor concentrations in the range of 50 to 5,000 ppmv (U.S. EPA, 1994, EPA/542-R-94-003).

Saberiyan et al. (1992) studied the use of a biofilter for treatment of air containing gasoline vapors. Sphagnum moss was used as the packing material. The system initially was inoculated with a hydrocarbon-degrading bacterial culture, then exposed to gasoline vapors. Up to 90% of the initial 50 ppmv gasoline vapor concentration was removed by the biofilter. These studies also sought to demonstrate the linear relationship between flowrate and packing material volume.

Biofiltration of vapor streams is a fairly well established treatment technology in Europe (Leson and Winer, 1991). Medina et al. (1995) have studied the use of biofilters to treat toluene and gasoline vapor streams. Bench- and pilot-scale reactors were studied.

Adsorption on Carbon or Resin

Adsorption refers to the process whereby molecules collect on and adhere to the surface of an adsorbent solid (U.S. EPA, 1988, EPA/530/UST-88/001). This adsorption is due to chemical and/or physical forces. Physical adsorption (the more common type in this application) is due to Van der Waals' forces, which are common to all matter and result from the motion of electrons. Surface area is the critical factor in the adsorption process because the adsorption capacity is proportional to surface area. Commercially available adsorbents include activated carbon and synthetic resins.

GAC is the most commonly used vapor phase treatment method. Because carbon has a complex internal pore structure, activated carbon adsorbents provide a high surface area in a low unit cost material. Commercially available GAC typically has a surface area of 1,000 to 1,400 m^2/gram.

GAC is the most cost-effective organic vapor treatment method for a wide range of applications due to its relative ease of implementation and operation, its established performance history in commercial applications, its ability to be regenerated for repeated use, and its applicability to a wide range of contaminants at a wide range of flow rates. Many vendors sell or lease prefabricated, skid-mounted units that can be put into operation with a few days notice. However, carbon adsorption is economical only for lower mass removal rates. When the vapor concentration is high, carbon replacement or regeneration may be prohibitively expensive.

On-site regeneration of the carbon is an alternative to carbon replacement with off-site disposal or reactivation. Such systems regenerate the carbon in place, using steam or hot gas to desorb the contaminants. The contaminants recovered in liquid form then can be disposed of or, in some cases, recovered as solvent or used as fuel.

Information on GAC design parameters is available from the carbon vendors. Calgon Carbon Corporation (Pittsburgh, PA), Carbtrol Corporation (Westport, CT), and Nucon (Columbus, OH) among many others, supply adsorption isotherms and pressure drop curves for the various GAC types they supply. The pressure drop curves are developed as a function of flow rate. Many vendors supply modular, prefabricated GAC units of 200 to 2,000 lb of activated carbon that may accommodate flow rates from below 400 standard cubic feet per minute (scfm) to more than 1,000 scfm.

As a rule of thumb, the adsorptive capacity of activated carbon for most hydrocarbons in the vapor stream is about 1 lb hydrocarbon:10 lb activated carbon, and the cost of activated carbon is about $3/lb (all costs included, not just carbon purchase, in 1993 dollars). Therefore, the cost of activated carbon treatment can be roughly estimated as being about $30/lb of hydrocarbon to be treated.

Specialized resin adsorbents have been developed and are now entering commercial application for treatment of organic vapors in off-gas streams. These synthetic resin adsorbents have a high tolerance to water vapor. Air streams with relative humidities greater than 90% can be processed with little reduction on the adsorption efficiency for organic contaminants. The resin adsorbents as amenable to regeneration on site. Skid mounted modules are available consisting of two resin adsorbent beds. The design allows one bed to be online treating off-gas while the other bed is being regenerated. During the desorption cycle, all of the organic contaminants trapped on the resin are removed, condensed, and transferred to a storage tank. The desorption process used to regenerate the resin is carried out under vacuum using a minimum volume of nitrogen purge gas. A heat exchanger in the bed heats the resin during regeneration. The same heat exchanger is used to cool the bed to increase sorption capacity while it is on line treating off-gas (Downey et al., 1994).

Catalytic Oxidation

Catalytic oxidation is an incineration process that uses catalysts to increase the oxidation rate of organic contaminants, yielding equivalent destruction efficiency at a lower temperature than for flame incineration. In catalytic oxidation, the vapor stream is heated and passed through a combustion unit where the gas stream contacts the catalyst. The catalyst accelerates the chemical reaction without undergoing a chemical change itself. The catalyst increases the oxidation reaction rate by adsorbing the contaminant molecules on the catalyst surface. Sorption phenomena on the catalyst serve to increase the local concentration of organic contaminants at the catalyst surface and, for some organic contaminants, reduce the activation energy required for the oxidation reaction. Increased concentration and reduced activation energy increase the rate of organics oxidation (Kiang, 1988). Figure D-2 shows a schematic diagram of a catalytic incinerator unit.

The active catalytic material is typically a precious metal (e.g., palladium or platinum) that provides the surface conditions needed to facilitate the transformation of the contaminant molecules into carbon dioxide and water. The catalyst metal is supported on a lower cost, high surface area metallic or ceramic support media.

The metal catalyst and support are exposed to the heated off-gas in a catalytic oxidation unit. The catalytic oxidation unit will use either a fixed bed or fluidized bed system. Fixed bed systems include metallic mesh, wire, or ribbon or ceramic honeycomb supporting the catalyst metal or a packed bed of catalyst-impregnated pellets. Fluidized beds also use catalyst-impregnated ceramic pellets but operate at sufficiently high flow to move and mix the pellets during treatment (U.S. EPA, 1986, EPA/625/6-86/014).

The main advantage of catalytic oxidation versus thermal incineration is the much lower temperature required with a catalyst. These systems typically operate at 600 to 900°F (CSM Systems, 1989), versus temperatures of 1,400°F or higher for flame incineration. The lower temperature results in lower fuel costs, less severe service conditions for the incinerator materials of construction, and reduced NO_x production. Natural gas or propane is a typical fuel used for supplemental heating when the contaminated vapor streams do not contain sufficient heat value for a self-sustaining incineration. Energy costs can be further reduced by reclaiming heat from the exhaust gases, i.e., using the exhaust gas flow to preheat the influent vapor stream.

Careful monitoring of extraction gas concentration and reactor temperature is required to prevent overheating of the catalyst. Overheating can damage the catalyst metal surface and/or the support reducing catalytic activity. The allowed influent organic vapor concentration depends on the heat value and LEL of the influent vapor stream. Concentrations over 3,000 ppmv VOCs normally are diluted with air to prevent excessive energy release rates and to control the temperature in the catalytic unit. Safety also is a concern with these units, as with any incineration method. The maximum permissible total hydrocarbon concentration varies by site but usually is below 25% of the LEL. The total hydrocarbon concentration in the vapor is measured continuously at the inlet to the catalytic unit to control the dilution air flow.

228

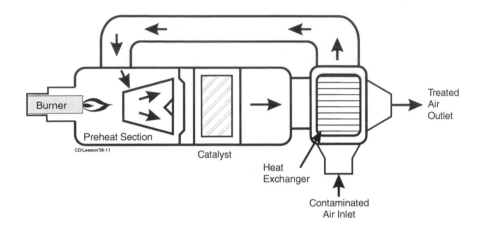

Figure D-2. Schematic Diagram of Catalytic Oxidation

Treating off-gas containing chlorinated compounds, sulfur-containing compounds, or nitrogen-containing compounds will deactivate the catalyst by chemical reaction of the catalyst metal with halogens or strong sorption of SO_x and NO_x on the catalyst. Some catalysts are specially designed for treatment of chlorinated compounds. New technologies potentially capable of treating chlorinated compounds by catalytic oxidation currently are under development and are beginning to appear on the market (Trowbridge and Malot, 1990; Buck and Hauck, 1992).

The significant cost elements of a catalytic oxidizer are the capital cost (or rental) of the unit; operations and monitoring; maintenance; and makeup fuel cost. A catalytic oxidation unit for treatment of 100 cfm off-gas flow would have a capital cost in the approximate range of $40,000 to $60,000 (in 1991 dollars) (AFCEE, 1992). Operations, maintenance, and monitoring costs will be site-specific. Makeup fuel will be required if the hydrocarbon concentration falls below the level necessary to maintain the required temperature. At the Hill Air Force Base 914 site (Smith et al., 1991), the extracted hydrocarbon concentration was approximately 600 ppmv and flow rate 1,500 cfm. To maintain the minimum temperature, an average of 1,500 gallons of propane was used every month at an average cost of $2,000 per month. All thermal oxidation processes will require makeup fuel to treat low concentration wastestreams, and the makeup fuel generally will be proportional to the operating temperature. Some fuel may be saved by heat recuperation.

Flame Incineration

Flame incineration uses high-temperature, direct-flame combustion to produce rapid oxidation of organic contaminants. Flame incinerators for treating organic vapors in off-gas are typically single-chamber, refractory-lined units containing an open burner. Flame incinerators often are equipped with heat exchangers, where hot combustion gases leaving the incinerator preheat the incoming off-gas stream. Heat recovered from the combustion gas improves thermal efficiency and reduces fuel costs. When operated with adequate temperature and residence time, flame incineration treatment will oxidize hydrocarbon contaminants to carbon dioxide and water. For most contaminants, acceptable contaminant destruction efficiency is achieved with an operating temperature in the range of 1,400 to 1,600 °F and a residence time of 1 second (AFCEE, 1992). Addition of makeup fuel is usually needed to maintain the temperature required to ensure adequate mineralization. Natural gas or propane typically serves as the supplemental fuel. The destruction of the contaminants is a major advantage of this technique over carbon adsorption, which only concentrates the contaminants onto the carbon, which must then be regenerated or disposed.

Safety is a major design requirement for flame incinerators and other thermal destruction units. Requirements for safety provisions are derived from National Fire Protection Association (NFPA) standards and applicable State requirements. In most applications, influent concentrations are limited to 25% of the LEL (AFCEE, 1992). The LEL for gasoline is between 12,000 ppmv and 15,000 ppm, depending on the gasoline's grade (Little, 1987).

Direct incineration is not appropriate for influent vapor streams that contain chlorinated compounds. Complete combustion of these compounds will generate corrosive hydrochloric acid vapors. Partial or incomplete combustion of chlorinated compounds could result in the production of chlorinated products.

The capital cost of a flame incinerator typically is less than that of a catalytic incinerator. However, due to the higher operating temperature, fuel use will be higher in a flame incinerator. When the flammable contaminant vapor concentration is sufficiently high, the heating value from oxidation of the contaminant reduces fuel use so at higher hydrocarbon concentrations flame incineration may be less costly than catalytic incineration. At lower vapor concentrations, the cost of makeup fuel will be much greater than for catalytic incineration, and the overall cost will probably be higher than for catalytic incineration. Flame incineration is generally favored over catalytic oxidation when the combustible organic vapor concentration is higher than about 1,000 to 5,000 ppmv (AFCEE, 1992).

230

Internal Combustion Engines

Internal combustion engine treatment accomplishes destruction of organic contaminants by oxidation in a conventional engine. ICEs have been used for years to destroy landfill gas. Application of ICEs to destruction of hydrocarbon vapors in air streams is more recent. The first operational unit was installed in 1986.

The ICE used for this technique is an ordinary industrial or automotive engine with its carburetor modified to accept vapors rather than liquid fuel. The air flow capacity of the ICE is determined by the cubic-inch displacement of the engine, the engine speed, and the engine vacuum. The capacity (scfm) can be estimated as follows:

$$\text{Capacity} = \frac{\text{RPM}}{2} \times \frac{\text{CID}}{1,728} \times 0.85 \times \left(1 - \frac{\text{EV}}{\text{P}}\right) \qquad \text{(D-1)}$$

where: RPM = engine speed in revolutions per minute
CID = engine displacement in cubic inches
EV = vacuum at the engine intake in inches of mercury
P = local air pressure in inches of mercury

Therefore, a 140-cubic-inch-displacement 4-cylinder engine running at 2,250 rpm and 10 inches Hg engine vacuum with an atmospheric pressure of 30 inches of mercury would have an off-gas treatment capacity of 52 scfm. ICE treatment units are available in sizes from 140 cubic inches to 460 cubic inches. Currently available ICE treatment units operate the engine at near-idle conditions. The off-gas capacity could be increased by applying a load to the engine to increase engine speed and decrease engine vacuum. Engine loading by attaching a generator to supply site power has been proposed but is not routinely practiced.

A second required modification to the engines is the addition of a supplemental fuel input valve when the intake hydrocarbon concentration is too low to sustain engine operation. Propane is used almost universally, although one vendor reported that natural gas, when available, can reduce energy cost by 50 to 75%.

The engines are equipped with a valve to bleed in ambient air to maintain the required oxygen concentration. Soil vapor may have very low concentrations of oxygen, especially during the initial stages of operation. Ambient air is added to the engine, via an intake valve, at a ratio sufficient to bring the oxygen content up to the stoichiometric requirement for combustion.

A catalytic converter is an integral component of the system, providing an important polishing step to reach the low discharge levels required by many regulatory agencies. A standard automobile catalytic converter, using a platinum-based catalyst, is normally used. Data from the South Coast Air Quality Management District, the air quality regulatory body for Los Angeles and the surrounding area, show that the catalyst reduced concentrations of TPH from 478 ppmv to 89 ppmv and from 1,250 ppm to 39 ppm, resulting in important additional contaminant removal (U.S. EPA, 1991, EPA/540/2-91/003). The South Coast Air Quality Management District requires a catalytic converter to permit this type of system. Catalysts have a finite life span (typically expressed in hours of operation) and must be monitored as the end of that time approaches to ensure that the catalyst is working properly. The length of operation of the catalyst depends on the vapor concentration and whether lead or other potential catalyst poisons are present in the off-gas contaminants. A range suggested by one equipment vendor was 750 hours to 1,500 hours (about 1 to 2 months) of operation. A deactivated catalyst can be replaced easily with any automobile catalytic converter, available at most auto parts stores.

To date, ICE use appears to be most widespread in California, mostly in the South Coast Air Quality Management District in southern California, which has among the most stringent air discharge regulations in the country. The South Coast Air Quality Management District has permitted more than 100 ICEs for use in their district. RSI, Inc. (Oxnard, CA) has installed more than 30 ICE systems, all in California.

Data obtained from ICE operators and regulators show that ICEs are capable of destruction efficiencies of well over 99% (U.S. EPA, 1991, EPA/540/2-91/003, p. 93). ICEs are especially useful for treating vapor streams with high concentrations of TPH (up to 30% volume) to levels below 50 ppm. Tests of benzene, toluene, ethylbenzene, and xylenes (BTEX) destruction by ICE treatment show that nondetectable levels of contaminants can be achieved in the outlet off-gas in some cases and outlet concentrations below 1 ppmv can be achieved in many cases. The total destruction capacity can be expressed as mass removal rate. One ICE operator reported a mass removal and destruction rate of over 1 ton per day (about 12 gph).

ICE off-gas treatment units are able to handle high concentrations of organic contaminants in the extracted air. As discussed above, incineration units, (e.g., catalytic oxidation units and flame incinerators) usually are limited to inlet vapor concentrations of 25% of the LEL. Because the inlet concentration for an ICE unit can be in the combustible range, these units can accept vapor concentrations as high as 40,000 ppmv with no dilution air. As a result, the ICE treatment units have a significant advantage over incineration units when the vapor concentration is higher than 25% of the LEL. Inlet vapor concentration as high as 300,000 ppmv have been reported (U.S. EPA, 1991, EPA/540/2-91/003). The off-gas must still be diluted with air to allow the ICE unit to treat off-gas containing more than about 40,000 ppmv of organics, but only one-quarter as much dilution air flow is needed for the ICE unit compared to an incineration unit.

ICEs also can effectively treat low concentrations (i.e., inlet vapor concentration below 1,000 ppm), although supplemental fuel use increases as the inlet concentration drops below 14,000 ppmv and the cost-effectiveness decreases at reduced intake concentrations. The removal efficiency compares favorably with other treatment methods, based on data available from actual system installations.

ICEs as vapor treatment devices for extracted soil vapors offer advantages over conventional treatment methods (carbon, thermal oxidation, or catalytic oxidation), at least for some applications. One advantage of ICEs is the ability to produce power that can provide useful work output. Self-contained units are available that use the ICE to power the blower. The extraction blower consumes only about 25% of the useful work produced by the engine. Other uses of the power have included lighting the site, heating a field trailer, or similar ideas. Using the engine as a vacuum source increases the engine vacuum, which has the undesirable effect of reducing the air flow capacity (see equation above). As a result, the ICE usually is coupled to a blower to supply the well head vacuum. An added benefit of this system is that vapors cannot be extracted unless treatment also is occurring, eliminating the possibility of vapors bypassing the treatment system.

Another advantage of ICEs is their portability and ease of monitoring and maintenance. Typically, the self-contained units are skid-mounted or put on a trailer and can go from site to site very easily. The site requirements may also favor ICEs over other oxidation methods. ICE units are smaller and less noticeable than direct thermal incineration units and may be more appropriate for areas that are intended to remain low profile. Units also have been developed that can be monitored via modem, eliminating costly on-site monitoring.

Noise associated with the operation of the engine could be a concern near residential areas or occupied buildings. Noise can be abated by adjusting engine speed during certain time periods, installing a noise-suppression fence, or purchasing special low noise ICE models (AFCEE, 1994).

The capital cost of currently available ICE units appears to be somewhat higher, but certainly in the same general range as for catalytic incineration and thermal incineration. The costs of ICE treatment units with maximum flow capacities of 65 scfm, 250 scfm, and 500 scfm are $40,450, $73,450, and $98,880 respectively (in 1994 dollars). Propane or natural gas fuel is needed when the inlet vapor concentration is below about 40,000 ppmv. The quantity of added fuel needed increases as the inlet vapor concentration declines. Fuel costs for treating 65 scfm, 250 scfm, and 500 scfm off-gas flow, when all energy is supplied by propane supplemental fuel, are $20/day, $70/day, and $140/day, respectively (AFCEE, 1994). Operations and maintenance costs are site-specific. Because ICEs use widely understood technology, gaining regulatory acceptance appears to be easier than for other technologies, and as a result, permitting and monitoring costs should be lower.

Emerging Vapor Treatment Technologies

This section briefly describes the operating features of several emerging technologies for destruction or concentration of organic contaminants in an off-gas stream. The technologies described are packed-bed thermal treatment, photocatalytic oxidation, and membrane separation.

Packed-bed thermal treatment oxidizes organic contaminants by passing the off-gas stream through a heated bed of ceramic beads. The packed bed increases mixing to promote oxidation. A vapor stream passes through the packed bed that thermally destroys the contaminants. The packing geometry combined with uniform high temperature of the ceramic beads is reported to give good destruction efficiency for organic vapors in air, without using an open flame. The ceramic beads are heated electrically to bring them to the operating temperature of 1800°F. No additional energy input is required if the heat value of the vapors is sufficient. This point is near a concentration of 2000 ppmv. If the concentration is below this value, natural gas or propane can be bled in with influent to maintain the proper temperature. As with any incineration technique, excess air is added to dilute the concentration to safe levels if the influent is too rich. Packed-bed thermal processing has been used to destroy vapor contaminants in the off-gas from several chemical and other industrial plants. The vendor of the technology currently is investigating its applicability to the remediation market (U.S. EPA, 1991, EPA/540/2-91/003).

The vendor indicates that this technology has several desirable characteristics for treatment of vapors in off-gas from remediation systems. The removal efficiency is reported to be high and stable over varying operating conditions. Tests have shown efficiencies of 99.99 + %, and this removal is attained continuously. Another reported advantage is the ability to mineralize chlorinated compounds without the production of chlorinated products of incomplete combustion or degradation of the ceramic beads. Mineral acid vapors still would be produced.

In the photocatalytic oxidation process, volatile organic compounds are converted to carbon dioxide and water by exposure to ultraviolet (UV) light. When chlorinated organics are present, hydrogen chloride gas and/or chlorine also are produced. The off-gas stream enters the photocatalyst unit where the contaminants are trapped on a catalyst surface. The catalyst surface is continuously exposed to high-intensity UV light. The combination of surface effects from the catalyst and energy input from the UV light allows rapid oxidation of the contaminants. The reported residence time required for 95 to 99% destruction efficiency is 0.2 seconds (Kittrel et al., 1995).

Gas semipermeable membrane systems are available to concentrate dilute organic vapor streams. The membrane systems do not destroy the organic contaminants and would, therefore, be used as a pretreatment step to increase the efficiency of a second treatment process. The membranes used have dramatically different permeability for air and organic vapor molecules. The difference in permeability allows the organics to concentrate on one side of the membrane and the air on the other side. The concentrated vapor stream can then be further processed to condense and collect the organics or destroy them (U.S. EPA, 1994, EPA/542-R-94-003).

REFERENCES

AFCEE. 1992. Remedial Technology Design, Performance, and Cost Study. Environmental Services Office, Air Force Center for Environmental Excellence, Brooks Air Force Base, TX.

AFCEE. 1994. A Performance and Cost Evaluation of Internal Combustion Engines for the Destruction of Hydrocarbon Vapors from Fuel-Contaminated Soils. Report prepared by S.R. Archabal and D.C. Downey, Environmental Services Office, Air Force Center for Environmental Excellence, Brooks Air Force Base, TX.

Buck, F.A.M, and C.W. Hauck. 1992, "Vapor Extraction and Catalytic Oxidation of Chlorinated VOCs." In: *Proceedings of the 11th Annual Incineration Conference.* Albuquerque, New Mexico, May.

CSM Systems. 1989. Company literature. Brooklyn, NY.

Downey, D.C., C.J. Pluhar, L.A. Dudus, P.G. Blystone, R.N. Miller, G.L. Lane, and S. Taffinder. 1994. "Remediation of Gasoline-Contaminated Soils Using Regenerative Resin Vapor Treatment and In Situ Bioventing." In: *Proceedings of the Petroleum Hydrocarbons and Organic Chemicals Ground Water: Prevention, Detection, and Restoration Conference.*

Kiang, Y-. H. 1988. "Catalytic Incineration." In: Harry Freemen, (Ed.), *Standard Handbook of Hazardous Waste Treatment and Disposal.* McGraw-Hill, New York, NY.

Kittrell, J.R., C.W. Quinlan, A. Gavaskar, B.C. Kim, M.H. Smith, and P.F. Carpenter. 1995. "Air Stripping Teams with Photocatalytic Technology in Successful Groundwater Remediation Demonstration." *Air and Waste Management Association Annual Meeting.* Paper number A469. San Antonio, Texas.

Leson, G. and A.M. Winer. 1991. "Biofiltration: An Innovative Air Pollution Control Technology for VOC Emissions." *Journal Air Waste Management Assoc. 41*(8), August 1991.

Little, A.D. 1987. The Installation Restoration Program Toxicology Guide, Vol. 3. Prepared for the Aerospace Medical Division, U.S. Air Force Systems Command, Wright-Patterson Air Force Base, Dayton, OH.

Medina, V.F., J.S. Devinny, and M. Ramaratnam. 1995. "Biofiltration of Toluene Vapors in a Carbon-Medium Biofilter." In: R.E. Hinchee, G.D. Sayles, and R.S. Skeen (Eds.), *Biological Unit Processes for Hazardous Waste Treatment.* Battelle Press, Columbus, OH, pp. 257-264.

Miller, R.N., C.C. Vogel, and R.E. Hinchee. 1991. "A Field-Scale Investigation of Petroleum Hydrocarbon Biodegradation in the Vadose Zone Enhanced by Soil Venting at Tyndall AFB, Florida." In: R.E. Hinchee and R.F. Olfenbuttel (Eds.), *In Situ Bioreclamation.* Butterworth-Heinemann, Stoneham, MA. pp. 283-302.

Saberiyan, A.G., M.S. Wilson, E.O. Roe, J.S. Andrilenas, C.T. Esler, G.H. Kise, and P.E. Reith. 1992. "Removal of Gasoline Volatile Organic Compounds Via Air Biofiltration: A Technique for Treating Secondary Air Emissions From Vapor-Extraction and Air-Stripping Systems." in R.E. Hinchee, B.C. Alleman, R.E. Hoeppel, and R.N. Miller (Eds.), *Hydrocarbon Bioremediation.* Lewis Publishers, Boca Raton, FL. pp. 1-12.

Skladany, G.J., A.P. Togna, and Y.-H. Yang. 1994. "Using Biofiltration to Treat VOCs and Odors." *Superfund XIV Conference and Exhibition.* Hazardous Materials Control Resources Institute, Rockville, MD.

Smith, G.J., R.R. Dupont, and R.E. Hinchee. 1991. Final Report on Hill AFB JP-4 Site (Building 914) Remediation. Report submitted to Hill AFB, Utah by Battelle. July.

Trowbridge, B.E. and J.J. Malot. 1990. "Soil Remediation and Free Product Removal Using In-Situ Vacuum Extraction with Catalytic Oxidation." In: *Proceedings of the Fourth Annual Outdoor Action Con-*

236

ference on Aquifer Restoration, Ground Water Monitoring and Geophysical Methods. Las Vegas, NV.

U.S. Environmental Protection Agency. 1986. *Control Technologies for Hazardous Air Pollutants*. EPA/625/6-86/014.

U.S. Environmental Protection Agency. 1988. *Cleanup of Releases from Petroleum USTs: Selected Technologies*. EPA/530/UST-88/001.

U.S. Environmental Protection Agency. 1991. *Soil Vapor Extraction Technology: Reference Handbook*. EPA/540/2-91/003. Office of Research and Development, Risk Reduction Engineering Laboratory, Cincinnati, OH.

U.S. Environmental Protection Agency, 1994. *VISITT - Vendor Information System for Innovative Treatment Technologies*. EPA/542-R-94-003. Solid Waste and Emergency Response, Washington, DC.

INDEX

237